Bioethanol and Natural Resources

Bioethanol and Natural Resources

Substrates, Chemistry and Engineered Systems

Ruben Michael Ceballos

CRC Press
Taylor & Francis Group
Boca Raton London New York

CRC Press is an imprint of the
Taylor & Francis Group, an **informa** business

CRC Press
Taylor & Francis Group
6000 Broken Sound Parkway NW, Suite 300
Boca Raton, FL 33487-2742

First issued in paperback 2020

ISBN: 978-0-367-57265-5 (pbk)
ISBN: 978-1-4987-7041-5 (hbk)

Library of Congress Cataloging-in-Publication Data

Names: Ceballos, Ruben Michael, author.
Title: Bioethanol and natural resources : substrates, chemistry and
engineered systems / Ruben Michael Ceballos.
Description: Boca Raton : CRC Press, [2017] | Includes bibliographical
references and index.
Identifiers: LCCN 2017022730 | ISBN 9781498770415 (hardback : alk. paper) |
ISBN 9781315154299 (ebook)
Subjects: LCSH: Cellulosic ethanol. | Biomass energy. | Biodiesel fuels.
Classification: LCC TP339 .C43 2017 | DDC 668.4/4--dc23
LC record available at https://lccn.loc.gov/2017022730

Visit the Taylor & Francis Web site at
http://www.taylorandfrancis.com

and the CRC Press Web site at
http://www.crcpress.com

Dedication

There is nothing wrong with making a little money. People who work hard should see the benefits of their individual efforts. However, when profit is gained off the backs of others or at the expense of our environment, then this is a problem. I was once told that things in this world [energy and matter] can neither be created nor destroyed, they can only be transformed from one state to another. I hear people talk about "creating wealth" in our society and it makes me think. It does not seem that wealth can be created from nothing. Generating what they call "new wealth" comes at the expense of something— either at the expense of a person's sweat and time (and thus his life) or at the expense of depleting something in nature.

It seems that a few people make huge profits at the expense of the hard work and lives of others who see only a very small fraction of the benefit that their efforts produce. It seems that these same few people also make huge profits off of the land and other natural resources without care for replacing these natural resources at a rate that matches the rate at which the resource is being depleted. This is not "creation of wealth" but the abuse of our fellow man and the abuse of our Earth. So, be aware as you move forward and seek profit. Make sure that the profit that you receive is commensurate with the work that you, yourself, have done and that you are not profiting at the expense of the lives of your neighbor or to the detriment of our planet.

Go forward and be a responsible, caring, and hard-working citizen of the world, not just for the sake of yourself, your family, and your people but for the benefit of mankind as a whole.

If you do this, then when you are an old man, I think you will look back on your life and feel good—that you have done the right thing. This is more important than any wealth that you may acquire.

This book is dedicated to Catarino Ceballos (Ódami/Tepehuano), my grandfather, who left school at the age of thirteen to work in the fields and then later on the Santa Fe railroad, so that he could buy school uniforms for his younger sisters and who, on retiring from the railroad, would enroll himself in Palomar College in San Marcos, California, to finish, as a senior citizen, the education that he had to leave behind as a young man. Both he and my grandmother, Isabel, always provided unwavering support for their children and grandchildren despite being of humble economic means.

As a descendant of both Sephardic and Native American ancestry and an avid seeker of knowledge, he never lost sight of the need for compassion for others nor the need to seek balance between modern development and sustainable use of our natural resources.

Catarino Ceballos (1916–1999)
Isabel Ceballos (1916–2001)

Contents

Preface

Research in alternative energy continues to be critical in science, engineering, and economics. From 2011 to 2017, the world population increased from an estimated 7.0–7.5 billion people. As the global population charges toward 8 billion (and beyond), the demand for liquid (or gas) fuel alternatives will exceed what is capable of being supplied by fossil-based fuel resources. We have already seen signs of this stress as fossil-fuel companies begin to extract crude oil using techniques such as *fracking* even though detrimental impacts on subsurface structures are obvious. Even more telling is the multitude of international and regional conflicts, including open war that arises from efforts to control regions with proven reservoirs of fossil fuel (i.e., crude oil). Although there are potential solutions to the continuing fuel crisis, controlling supply allows those who profit from fossil fuels to manipulate demand toward generating significant profit. Until this civilized world decides that this is no longer acceptable, alternative fuels will continue to be suppressed and the war and destruction of our natural resources will continue.

If at some point in human history, society decides that the loss of human life and the destruction or depletion of our natural resources in an unsustainable manner are unacceptable, a holistic approach that provides a diverse array of liquid fuel options may be adopted. It is also essential that production of alternative fuels should not produce detrimental side effects on the environment. Liquid fuel must be produced and used with limited and calculated impact on air quality, water quality and availability, food crop lands, geological stability, and other factors that are essential for life on this planet.

Although bioethanol, particularly *cellulosic ethanol*, is only one of several options to replace fossil-based liquid fuel, the potential to significantly contribute to the need for additional fuel products is great given that cellulosic biomass is widely abundant and often simply treated as waste from other operations (i.e., corn, rice, sugarcane farming). A relatively small group of researchers across the world quietly chug away at developing novel technologies to access usable energy molecules from lignocellulosic biomass.

The goal, of course, is to develop technologies and processes that reduce the per-gallon cost of a bioethanol product, so that it may someday reach the market in full force and genuinely compete with fossil-based liquid fuel.

Our lab is one of many that are dedicated to finding new ways to deconstruct lignocellulose in a cost-effective manner. We are a very small group compared to larger labs across the world; however, the information that we require to forward our technologies is the same as what anyone in this field would need. As this field moves quickly with new discoveries happening almost daily, it was important to get this book to press in a timely manner. There are likely corrections and additions that could be incorporated into a subsequent edition. We welcome suggestions from the community. It is hoped that this book will provide a somewhat comprehensive technical review of substrates, feedstocks, enzymes, approaches, and technologies that are available on the quest to bring bioethanol to the forefront of liquid fuel alternatives.

Acknowledgments

I thank Inés Cuesta Ureña who completed an extensive literature review in support of this book (and her master's thesis), which was used to develop selected sections of the original manuscript. I thank Natalia A. Batchenkova, an undergraduate student in my lab, who constructed tables, modified figures from other sources, and created several of the figures used in this book *de novo*. Ms. Batchenkova also provided a literature review in support of selected sections of this book.

Inés Cuesta Ureña
Master in Interdisciplinary Studies
University of Montana
(2014)

Natalia A. Batchenkova
Bachelor of Science (Biology)
University of Arkansas
(Class of 2018)

I also acknowledge the: (1) University of Arkansas Center for Space and Planetary Science (SPAC), Fayetteville, AR; (2) Minority Institution Astrobiology Collaborative (MIAC); (3) Native American Research Lab (NARL); (4) Biofuels Research Collaborative for using Microorganisms

As Renewable Energy Sources (BRC-MARES); and (5) Blue Marble Space Institute for Sciences (BMSIS), Seattle, WA—for forwarding work in Earth/ecosystem science, extremophile biology, and alternative energy.

I also thank Dr. Douglas Rhoads, Dr. David McNabb, Dr. Ralph Henry, Dr. Steve Beaupre, and Dr. Jeannine Durdik of the Department of Biological Sciences at the University of Arkansas, Fayetteville, Arkansas, who have provided strong support for my efforts at the university. I thank Dr. Rob Gardner, my colleague at BRC-MARES, for his collegiality and collaboration. I also must acknowledge the U.S. National Science Foundation for supporting my research and student training efforts over the years through multiple grants, including U.S. national science foundation research experiences for undergraduates program (NSF BIO REU) awards. This includes NSF award numbers: 1742602, 1624171, 1615544, 1521023, 1342631, 1359324, and 0929484. Finally, I thank Hilary LaFoe and the staff at Taylor & Francis Group for their patience during the preparation of the original manuscript for this book and I thank Toan Trung Nguyen for his artwork on the book cover.

About the author

Dr. Ruben Michael Ceballos is a faculty member in the Department of Biological Sciences in the J. William Fulbright College of Arts and Science at the University of Arkansas, Fayetteville, Arkansas. He received a bachelor of science degree in physics and mathematics from the University of Alabama in Huntsville and a master of arts degree in behavioral neuroscience from the University of Alabama at Birmingham. He earned his doctor of philosophy degree in integrative microbiology and biochemistry from the University of Montana in Missoula. Dr. Ceballos is an Alfred P. Sloan Indigenous Graduate Program PhD graduate, a Ford Foundation Dissertation Fellowship awardee, the 2006 U.S. National Aeronautics and Space Administration (NASA) minority institute research support (MIRS) awardee, and a graduate of a U.S. National Science Foundation Integrative Graduate Education and Research Traineeship (IGERT) "Montana Ecology of Infectious Disease" PhD program. He is an affiliate of the nonprofit Blue Marble Space Institute for Sciences and a steering committee member for the NASA Minority Institution Astrobiology Collaborative (MIAC). Along with Dr. Rob Gardner of the University of Minnesota, he cofounded the Biofuels Research Collaborative—Microorganisms As Renewable Energy Sources, a virtual collaboration of students and scientists who are dedicated to the developing of microbial-derived renewable energy technologies and processes.

As one of a handful of doctorate-prepared Native American scientists, Dr. Ceballos has a long history of training students from ethnic, racial, and economic groups who have been historically underrepresented in the sciences. The Ceballos Lab is a basic protein biochemistry and microbiology research facility that pursues interdisciplinary research projects ranging from studies of viral-induced epileptogenesis to the development of microbial-based biotechnology for the enhancement of biofuels production. The Ceballos Lab has extensive research and student training partnerships worldwide, with active collaborations in Vietnam, Thailand, Malaysia, India, Mexico, Colombia, Cuba, and Norway.

Introduction

The unsustainable use of natural resources and an increasing demand for energy are two major concerns that must be addressed on a global scale if modern lifestyles are to be continued. With respect to these concerns, dependency on fossil fuels is unsustainable due to an apparent inability to discover new sources of nonrenewable substrates at a pace that matches the increasing global consumption of energy products (Ragheb, 2015). This situation has prompted research into alternative energy products, which are considered *sustainable* and *environmentally friendly*. Although advances in hydroelectric, solar, wind, and biomass systems provide viable alternatives for electrical energy production, substitutes for liquid fossil fuels have been problematic. *Biofuels*, which are alternatives to liquid fossil fuels, or *transportation fuels*, can be made from carbohydrate- or lipid-rich feedstocks that yield bioethanol and biodiesel, respectively (Demirbas, 2008). The production and use of bioethanol as a supplemental transportation fuel (and a partial substitute) to gasoline is ongoing. Indeed, the alternative liquid fuels market is a trillion dollar per year industry with ~15 billion gallons per year of ethanol produced in the United States alone (U.S. Energy Information Administration, 2015). Virtually all of this is corn-based ethanol, which accounts for ~10% of the United States transportation fuel needs (Hill et al., 2006). Sugarcane-based ethanol and smaller quantities of biodiesel augment corn-based products to account for the majority of biofuel produced worldwide. However, there are significant limitations in the large-scale production of bioethanol that prevent it from becoming a cost-competitive, *bona fide* substitute for liquid fossil fuels. Primary limitations in bioethanol production include: (1) the inefficient conversion of feedstock molecular substrates (i.e., cellulose and starch) to fermentable sugars such as glucose (Sukumaran and Pandey, 2009) and (2) the inability to use cellulosic feedstocks as viable substitutes for corn and sugarcane (Somma et al., 2010). Given that cellulose is the most abundant biopolymer on Earth and that corn and sugarcane are needed for global food supplies, it is reasonable to suggest that scientific research and industrial interests should focus on overcoming these limitations, so that cellulosic bioethanol may become a cost-competitive alternative to liquid fossil fuels.

Of particular interest is the development of enzyme systems that can reduce the cost of producing *first-generation* alternative fuel products (i.e., corn- and sugarcane-based ethanol) and facilitate the cost-effective production of *second-generation* fuel (i.e., cellulosic ethanol). The evolutionary emergence of cellulose-degrading microbes provides opportunities for development of ethanol production processes that utilize naturally occurring enzymes. More recently, genetically engineered enzymes have been employed to enhance enzymatic efficiency and lower the costs associated with commercial bioethanol production processes (Mohanram et al., 2013). Both naturally derived and genetically engineered enzymes are used in first-generation biofuels (FGBs) production processes. In addition, several advances in enzyme system technology have paved the way for cost-competitive production of second-generation biofuels (SGBs) (Mitsuzawa et al., 2009).

The purpose of this book is to provide a comprehensive review of the various enzymes that are used in bioethanol production, an explanation of their respective modes of enzymatic action, and a description of the various ways that these enzymes may be employed to facilitate efficient reduction of complex biomolecular substrates to readily fermentable sugars. This book begins with a description of various feedstock and pretreatment alternatives and addresses the advantages and disadvantages of each feedstock and pretreatment alternative.

I begin with a review of the molecular substrates found in FGB and SGB feedstocks that are important to ethanol production processes. This is followed by a description (and comparison) of production potential, chemical composition, and pretreatments of several high-yield FGB and SGB feedstocks. Next, I introduce a series of enzymes involved in the breakdown of lignocellulosic materials with a particular focus on the mode of action of each class of enzyme. Subsequently, the breakdown of substrates by both natural and artificial multienzyme systems is addressed, including a discussion of synergistic effects. This is followed by a discussion of emerging technologies and engineered systems. Finally, I provide the summarizing remarks.

chapter one

Molecular substrates of ethanol feedstocks

Regardless of the type of feedstock, all bioethanol production processes rely on isolating one or more molecular substrates that are used to generate precursors for ethanol-producing reactions. These molecular substrates include starch, sucrose, cellulose, hemicellulose, lignin, and pectin—which may be further decomposed for use (or used directly) by yeast in ethanol fermentation. Whereas the role of yeast in ethanol fermentation is a well-studied component of the overall process, the conversion of these molecular substrates to readily fermentable sugars (i.e., glucose) is a subject of ongoing research and is often considered a *bottleneck* in bioethanol production. Indeed, the culturing of yeast and ethanol production are typically two tightly coupled processes used in fermentation via glycolysis, the Embden–Meyerhof–Parnas pathway (Madigan et al., 2000). Glycolysis ultimately yields two pyruvate molecules for each molecule of glucose that is processed. Under anaerobic conditions, pyruvate is subsequently reduced to ethanol with a release of CO_2. Thus, enhancing processes that decompose complex molecular substrates to glucose or other fermentable sugars is a key to increasing overall bioethanol production efficiency.

1.1 Starch and sucrose

Historically, the production of ethanol has relied almost exclusively on two molecular substrates: starch and sucrose from corn (*Zea mays* L.) and sugarcane (*Saccharum* L.), respectively. More broadly, starches from corn and wheat and reducing sugars from sugarcane and sorghum are extracted, typically via mechanical processes and serve as substrates for subsequent ethanol fermentation steps (Nichols and Bothast, 2008; Dias et al., 2009).

1.1.1 Starch

Corn is milled to extract starch. There are two types of corn milling in the industry: wet and dry. During *wet milling*, the corn grain is pulverized and then separated into its various components: starch, fiber, gluten, and germ. On separating the starch as a solution from other components, it is typically treated with enzymes to generate simpler sugars that are easily fermented by yeast. The remaining components of the wet-milling process are often sold separately as coproducts. In wet milling, grains are steeped in dilute sulfuric acid prior to separation of grain components. However, in *dry milling*, whole grain *meal* is directly fermented to produce ethanol and by-products collectively called distiller's dried grains with solubles (DDGS), the latter of which may be used for animal feed or extracted for oil, which in turn may be converted to biodiesel via trans-esterfication (Singh and Cheryan, 1998; Bothast and Schlicher, 2005).

Although initial treatment of corn grain and the types of by-products that are generated differ between the two processes, both wet and dry milling ultimately break down starch via enzymatic processes, convert sugar to ethanol releasing CO_2 using yeast, and refine/dehydrate the resulting ethanol for use as fuel. In corn-based bioethanol production, starch is an abundant and readily hydrolysable substrate. Indeed, starch is the major chemical component within the corn kernel. It accounts for ~72% of kernel weight with another ~2% consisting of sugars such as glucose, sucrose, and fructose (Inglett, 1970). Cornstarch is generally composed of two types of polymers: *amylose* and *amylopectin*. Amylose is essentially a linear polymer of glucose. Amylopectin is a branched glucose polymer.

Of the total starch content in corn kernel, about 20%–25% is amylose and 75%–80% is amylopectin (Schoch, 1942; Bates et al., 1943). However, yeast is not generally capable of metabolizing starch (Stewart and Russell, 1987; De Mot, 1990). Thus, amylose and amylopectin must be hydrolyzed. Several enzymes, including thermostable α-amylases, are used to break down starch into smaller molecules (i.e., dextrins), which are then enzymatically converted to glucose (a.k.a., dextrose) by various glucosidases (e.g., amyloglucosidase) in a process called saccharification (Bothast and Schlicher, 2005). As over 90% of ethanol production in the United States uses starch as a primary molecular substrate, the importance of thermostable and high-efficiency starch reducing enzymes is understood.

1.1.2 Sucrose

In contrast to cornstarch, sucrose is the principal molecular substrate in sugarcane-based bioethanol production. Sucrose is found in high concentrations in both cane juice and molasses (a by-product of sugar mills). It is the only major ethanol substrate that does not require industrial enzymatic processing. In other words, yeast can directly metabolize sucrose to produce ethanol. In this respect, sucrose has a major advantage as a molecular substrate in ethanol production.

Ethanol production using substrates from sugarcane begins with cleaning the feedstock followed by mechanical extraction and separation of juice, molasses, and bagasse. Sugarcane juice contains water, sucrose, and other reducing sugars. It also contains impurities such as minerals, salts, organic acids, dirt, and fiber particles, which must be removed before fermentation (Dias et al., 2009). Clarification and crystallization steps can be used to extract sugar crystals leaving behind molasses (Ensinas et al., 2009). For a 3 lb sugarcane stalk, approximately 0.3 lb of sugar is generated, and 100 lb of raw sugar yields about 3 gal of molasses. From each gallon of molasses, 0.41 gal of ethanol can be produced. Although both products can be used to produce ethanol, only the molasses is often used. However, if both the raw sugar and molasses are converted to ethanol, approximately 19.6 gal of ethanol can be produced per ton of harvested sugarcane (Shapouri et al., 2006).

Most industrial strains of yeast express invertases (e.g., sucrases) that convert sucrose to simple sugars (i.e., glucose and fructose), which are in turn readily converted to ethanol (and CO_2) (Berthelot, 1860; Dworschack and Wickerham, 1961; Bowski et al., 1971). Thus, none of the early processing (e.g., liquefaction) that is required for using cornstarch as a molecular substrate is required for ethanol production that uses sucrose as the principal substrate. This saves several steps in the ethanol production process. The latter stages of fermentation and ethanol recovery are similar in both sugarcane-based and corn-based bioethanol production. Distillation and dehydration are used to purify the ethanol following fermentation by yeast (Bothast and Schlicher, 2005; Ensinas et al., 2009), and the remaining stillage from the distillation process can be recycled as fertilizer for sugarcane fields. Although sucrose is a competitive molecular substrate for bioethanol production, the feedstock from which it is derived (i.e., sugarcane) has production limitations as described in Section 2.1.

1.2   Cellulosic substrates

One of the most promising molecular substrates for bioethanol production is lignocellulose. Cellulose is the most abundant biopolymer on the planet. It accounts for approximately half of all biomass contained within photosynthetic plant matter. However, lignocellulosic materials are often bound in very recalcitrant matrices of cellulose, hemicellulose, lignin, and other molecules such as pectin (Table 1.1), which together render these materials unviable as competitive substrates due to the costs associated with accessing and breaking down these components into fermentable sugars. Nonetheless, emerging technologies that overcome this technical problem may soon elevate lignocellulose to a major substrate for industrial scale ethanol production.

Table 1.1 Chemical composition of cellulosic biomass (% composition)

Constituents	Hardwood (%)	Softwood (%)
Cellulose	40–50	40–50
Hemicellulose	25–35	25–30
Lignin	20–25	25–35
Pectin	1–2	1–2
Starch	Trace	Trace

Source: Miller, R.B., Structure of wood, in *Wood Handbook: Wood as an Engineering Material*, USDA Forest Service, Forest Products Laboratory, Madison, WI, 113, 1999.

1.2.1 Cellulose

Cellulose comprises at least 30%–50% of the total biomass in most lignocellulosic materials (Wyman, 1994; Mielenz, 2001; Mood et al., 2013). Cellulose is a linear, *unbranched* homopolysaccharide composed of β-D-glucopyranose units linked by β-1,4 glycosidic bonds (Jorgenson, 1950; Mark and Tobolsky, 1950; Hon, 1994). Chemically, cellulose is simply a chain of glucose molecules. However, structurally, the repeating unit is the disaccharide cellobiose (Staudinger, 1961). Individual cellulose chains can contain 10^2 to 10^4 glucose units. These chains are tightly packed into microfibrils with extensive hydrogen bonding as well as van der Waals interactions within and between cellulose chains, which provides stability to this high-order fibrous structure (Marchessault and Sarko, 1967; Cousins and Brown, 1995; Nishiyam et al., 2010).

Depending on the source and conditions of formation, cellulose will often form a crystalline structure; however, amorphous cellulose may also form a crystalline structure. With respect to the former, there are six distinct crystalline forms. Cellulose I (natural cellulose) and cellulose II (regenerated and mercerized cellulose) have been the most extensively studied (Habibi et al., 2010). Although crystalline cellulose can be difficult to degrade, amorphous cellulose tends to be more susceptible to enzymatic degradation (Hall et al., 2010). In either form, natural materials containing cellulose typically contain other polymers including hemicellulose, lignin, and pectin, which complicate the deconstruction of cellulose into glucose or other readily fermentable sugars.

1.2.2 Hemicellulose

Hemicellulose comprises approximately 15%–35% of the total biomass in lignocellulosic materials (Wyman, 1994; Gírio et al., 2010; Mood et al., 2013). Hemicellulose is a branched heteropolymer of hexose sugars, pentose sugars, and sugar acids. Of these constituent components, D-mannose, D-glucose, and D-galactose are common hexoses; D-xylose and L-arabinose are common pentoses; and D-glucuronic acid, D-galacturonic acid, 4-O-methylglucaronic acid, and rhamnose are common sugar acids found within hemicellulose. Sugars are linked by β-1,4 bonds and β-1,3 glycosidic bonds. The former dominates most hemicellulosic structures with the main backbone typically composed of only one or two types of sugar. For example, xylose, in the form of 1,4-β-D-xylopyranose structural units, forms the homopolymeric backbone of structures called xylans. As significant branching from this backbone occurs (typically consisting of short chains of other sugars, acetyl groups, or phenolic groups), most xylans are ultimately heteropolymers. Xylans can be classified as linear xylans, heteroxylans, or xyloglucans (XGs). Linear xylans consist of 1,4-β-D-xylopyranose units linked to form a polymeric backbone. However, xylans often contain additional components including arabinose, glucuronic acid, feruloyl acid, or coumaroyl acid side chains and are thus heteroxylans. Based on backbone and branching motifs, heteroxylans can be subclassified as arabinoxylans, glucuronoxylans, and other combinatory species, such as glucuronoarabinoxylans (GAXs) (Schulze, 1891; Scheller and Ulvskov, 2010). The frequency and composition of branches depend upon the source of xylan (Gorbacheva and Rodionova, 1977; Aspinall, 1980; Fincher and Stone, 1981; Brice and Morrison, 1982; Scheller and Ulvskov, 2010). In Figure 1.1, the heterogeneity of selected xylan structures is shown.

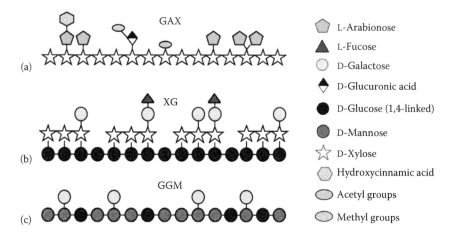

Figure 1.1 Plant cell wall polysaccharides and structural heterogeneity. (a) Backbone structure of heteroxylan, (b) Backbone structure of XG, and (c) Representative structure of a heteromannan molecule. (Adopted and modified from Burton, R.A. et al., *Nat. Chem. Biol.*, 6, 724–732, 2010.)

Acetyl groups and glucuronic acid moieties are found in hardwood hemicellulose (Timell, 1960; Bouveng, 1961; Timell, 1967; Gabrielii et al., 2000; Teleman et al., 2002). Arabinofuranosides and glucuronic acids are found in hemicellulose of softwoods and grasses (Marchessault et al., 1963; Timell, 1967; Shimizu et al., 1978; Wende and Fry, 1997; Verbruggen et al., 1998b; Pastell et al., 2009; Peng et al., 2011; Escalante et al., 2012). These heteroxylan (GAX) motifs are illustrated in Figure 1.1a. XG is a cellulosic (1,4)-β-glucan backbone supporting α-D-xylose side chains with about 70% occupancy of the backbone glucosyl residues linked at the glucose C-6 position. Triplet xylosyl substituents on three consecutive glucosyl residues are common along with additional side chain components such as galactose, fucose, or arabinose (Stephen, 1989; Sims et al., 1996; Hoffman et al., 2005; Tiné et al., 2006). A representative XG is shown in Figure 1.1a (side chain arabinose not shown).

Although xylans are the major components of hemicellulose, the heterogeneity of hemicelluloses should not be understated. Hemicelluloses are mixtures of polysaccharides and often include other components. Although hardwoods, grasses, and many fruits feature a robust presence of xylans, mannans play a major role in the hemicellulose found in softwoods and plant seed (Wilkie, 1979; Meier and Reid, 1982; Vierhuis et al., 2000; Lundqvist et al., 2003; Gírio et al., 2010). Mannans can be classified as linear mannan (e.g., glucomannan) or heteromannan (i.e., galactoglucomannan [GGM]). Glucomannan are linear chains composed of randomly arranged β-(1,4)-linked D-mannose with β-(1,4)-linked D-glucose. The ratios of mannose and glucose depend on the origin of glucomannan (Timell, 1967; Northcote, 1972; Popa and Spiridon, 1998; Hongshu et al., 2002). Another form of linear mannan called galactomannan (GM) features 1,6-linked α-D-galactopyranosyl side chains along the standard 1,4-linked β-D-mannopyranosyl main chain (Chaubey and Kapoor, 2001; Prajapati et al., 2013).

Heteromannan, or GGM, incorporates the β-(1,4)-linked D-glucose into the 1,4-linked β-D-mannopyranosyl main chain as well as the 1,6-linked α-D-galactopyranosyl and β-D-galactopyranosyl single-group side chains (Aspinall et al., 1962; Popa and Spiridon, 1998; Willfor et al., 2003; Hannuksela and du Penhoat, 2004). The galactose units are attached to either mannose or both mannose and glucose residues in the backbone (Aspinall et al., 1962; Matheson, 1990; Popa and Spiridon, 1998; Willfor et al., 2003; Hannuksela and du Penhoat, 2004). A representation of GGM is illustrated in Figure 1.1c.

The presence of xylans and mannans in lignocellulosic feedstock selection is important when considering feedstock pretreatment and enzyme-mediated systems for substrate deconstruction. Whereas GGM is a key component of softwood hemicelluloses, xylans such as glucuronoxylans are predominantly found in hardwood hemicelluloses (Timell, 1967;

Sjöström, 1993; Rowell et al., 2012). Unlike the crystalline structure of cellulose, hemicellulose forms a gel-like consistency that results in an amorphous matrix within which rigid crystalline cellulose fibers are often embedded. This facilitates the separation of fibers while adding flexion to the cell wall (Fratzl et al., 2004). Moreover, hemicelluolose can be linked to lignin yielding an even more complex composite with added structural stability (Das et al., 1984).

1.2.3 Lignin

Lignin comprises approximately 15%–20% of the total biomass in lignocellulosic materials (Wyman, 1994; Mood et al., 2013). Unlike cellulose and hemicellulose, lignin is a noncarbohydrate aromatic heteropolymer formed through oxidation and free-radical coupling of phenyl alcohol precursors (Boerjan et al., 2003; Ralph et al., 2004). Lignin macromolecules are comprised of phenolic (10%–20%) and nonphenolic (80%–90%) moieties with the latter exhibiting highly recalcitrant properties (Adler, 1977). The heterogeneous three-dimensional network of linked phenolic and nonphenolic structures formed by lignin within lignocellulosic materials may consist of dimethoxylated (syringyl, S), monomethoxylated (guaiacyl, G), and non-methoxylated (*p*-hydroxyphenyl, H) phenylpropanoid monomeric units (a.k.a., monolignols) derived from associated *p*-hydroxycinnamyl alcohols. These phenylpropanoid units are typically joined by nonhydrolyzable linkages, including covalent ether and carbon–carbon bonds (Ralph et al., 2004). Consistent with variations in the amount and type of lignin found in different raw materials, proportions of the different monolignols also vary between sources. For example, softwoods are known to contain a higher percentage of lignin than hardwoods, and softwood lignin is dominated by guaiacyl units, whereas hardwoods have approximately equal amounts of guaiacyl and syringyl units (Sjöström, 1993; Brunow, 2006; Santos et al., 2012; Normark et al., 2014). Thus, primary composition and structure of lignin are determined by species-specific genetics. Moreover, variation in amount and type of lignin is also observed between different tissues of the same plant. In the context of biofuel feedstock selection, it should be noted that the extraction method employed to isolate lignin from lignocellulosic biomass may confound efforts to elucidate lignin composition and structure in a natural living system (Gosselink et al., 2004).

Lignin is not water soluble, and the amorphous structure that it forms lacks the stereoregularity characteristic of cellulose (Hatakeyama and Hatakeyama, 1982). Indeed, within a composite lignocellulosic structure, lignin can often form bonds with both hemicellulose and cellulose through covalent cross-linking (Bell and Wright, 1950; Gerasimowicz et al., 1984; Balakshin et al., 2011). In contrast to hemicellulose and cellulose, lignin is not susceptible to hydrolytic attack, which renders most lignocellulosic materials resistant to biodegradation (Li et al., 2008a; Saritha and Lata, 2012). Lignin contributes to the structural support and cell envelope impermeability that many plants require to be optimally fit in their respective ecosystems. Lignin also contributes to a plant's resistance to microbial attack as well as oxidative stressors (Iiyama et al., 1994). Collectively, the amorphous nature of lignin, its lack of solubility in water, and optical inactivity are ideal for plant survival; however, as a component in biofuel feedstocks, these same properties are what make lignocellulosic materials difficult to degrade (Hatakeyama and Hatakeyama, 1982, 2010; Ralph et al., 1999). It is generally accepted that lignin depolymerization is the key to readily accessing cellulose and hemicellulose for efficient enzymatic degradation of lignocellulosic materials.

1.2.4   Pectin

Pectin comprises approximately <10% of the total biomass in selected lignocellulosic materials (O'Neill and York, 2003). Pectin-rich biomass generated as waste products from industrial processing of fruits and vegetables has low lignin and high pectin concentrations (Table 1.2), ranging from 12% to 35% of dry weight biomass (Edwards and Doran-Peterson, 2012) (Figure 1.2). Pectins are a family of covalently linked galacturonic acid-rich polysaccharides found in plant cell walls (Albersheim et al., 1996). Galacturonic acid comprises at least 65% of pectin and typically features high levels of methyl esterification (Phatak et al., 1988; Food & Agriculture Organization of the United Nations, 2009; Yapo and Koffi, 2014). All pectin polysaccharides contain galacturonic acid linkages at the O-1 and O-4 positions. Pectin polysaccharides are composed of four major subclasses: homogalacturonan (HG), rhamnogalacturonan I (RG-I), rhamnogalacturonan II (RG-II), and xylogalacturonan (XGA) (Yapo, 2011). HG is a simple structured linear homopolymer of α-1,4-linked galacturonic acid residues and comprises about 65% of pectin (Mohnen, 2008; Wang et al., 2012a). Galacturonic acid residues within this backbone can be methyl esterified at the carboxyl groups at the C-6 position (Gee et al., 1959; Mort et al., 1993; Petersen et al., 2008) and/or O-acetylated at the hydroxyl groups at the O-2 or O-3 positions (Perrone et al., 2002). The degree of methylation and acetylation varies significantly between sources (Wang et al., 2012a; Wang et al., 2014a). Other pectin polysaccharides are considerably more complex. RG-I represents approximately 20%–35% of all pectin. It is composed of a backbone of alternating rhamnose and galacturonic acid subunits that can consist of more than 100 repeating subunits (McNeil et al., 1980; Mohnen, 2008). The rhamnose molecules are often substituted with a variety of side chains mainly composed of arabinans, galactans, and arabinogalactans (Tharanathan et al., 1994; Nakamura et al., 2002; ØBro et al., 2004).

Table 1.2 Chemical composition (% dry wt.) of various lignocellulosic feedstocks

Feedstock type	Ash	Sugar	Fat	Protein	Pectin	Lignin	Cellulose	Hemicellulose
Mandarin peel	5.0 (3.0)	10.1	1.6	7.5	16.0 (14.2)	8.6 (8.9)	22.5 (20.8)	6.0 (17.2)
Lemon peel	2.5 (4.1)	6.5	1.5	7	13.0 (9.0)	7.6 (8.3)	23.1 (22.8)	8.1 (22.4)
Orange peel	2.6 (3.8)	9.6	4	9.1	23.0 (12.3)	7.5 (8.4)	37.1 (22.0)	11.0 (19.9)
Grapefruit peel	8.1 (3.3)	8.1	0.5	12.5	8.5 (16.1)	11.6 (8.2)	26.6 (19.8)	5.6 (18.3)
Soybean hulls	1.0–2.8	N/A	N/A	9.0–14.0	6.0–15.0	1.0–4.0	29.0–51.0	10.0–20.0
Corn stover	4.0–8.0	N/A	N/A	4.0–9.0	0	16.0–23.0	31.0–41.0	20.0–34.0
Wheat straw	1.0–10.0	N/A	N/A	2.0–6.0	0	5.0–19.0	32.0–49.0	23.0–39.0

Source: Boluda-Aguilar, M. and López-Gómez, A., *Ind. Crops Prod.*, 41, 188–197, 2013.

Note: N/A—data not available. Values in brackets represent data derived using TG–DTG method.

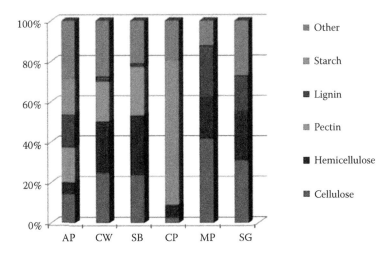

Figure 1.2 A comparison of the dry weight composition of pectin-rich biomass to starches and other lignocellulosic biomasses. Pectin-rich biomass includes apple pomace (AP), citrus waste (CW), and sugar beet pulp (SB). Corn kernel (CK), Monterey pine (MP), and switchgrass (SG) are included in other lignocellulosic biomasses. (Adopted from Edwards, M.C. and Doran-Peterson, J., *Appl. Microbiol. Biotechnol.*, 95, 565–575, 2012.)

The arabinogalactan polymeric forms can include side chains with type I arabinogalactan (AGI) and type II arabinogalactan (AGII) (Clarke et al., 1979). Feruloyl esters are attached to O-2 or O-6 of α-(1→5)-linked arabinose residues of the arabinan side chains or β-(1→4)-linked galactose in the galactan side chains of pectin, respectively (Guillon and Thibault, 1989; Guillon et al., 1989; Colquhoun et al., 1994; Ralet et al., 1994). RG-II is the most structurally complex pectic polysaccharide and accounts for 10% of all pectin. It consists of an HG backbone of at least 8 1,4-linked α-D-galacturonic acid residues decorated with side branches consisting of at least 12 different types of sugars (including rare sugars) that are covalently bound using over 20 different types of covalent linkages (O'Neill et al., 2004). XGA is an HG with a linked xylose at the O-3 position (Bouveng, 1965). With less prevalence, a xylose may also be linked at the O-4 position with an additional β-linked xylose (Zandleven et al., 2006). The degree of xylosylation can vary between 25% in fruits such as *Citrus vulgaris* (watermelon) and approximately 70% in *Malus domestica* (apples) and pea hulls (Schols et al., 1995; Yu and Mort, 1996; Goff et al., 2001). General distribution of methyl esters is typically across the XGA backbone (Schols et al., 1995).

1.2.5   Cross-linking: Hemicellulose

All plant-derived cellulosic materials in nature are a combination of cellulose and hemicellulose with various amounts of lignin, pectin, and other molecules contributing to the composite. Interestingly, the two principal components—cellulose and hemicellulose—do not readily associate via covalent bonding. Instead, cross-linking between cellulose and hemicellulose is dominated by hydrogen bonding. For example, XG will readily form hydrogen bonds with cellulose microfibrils in a manner that resembles interactions between cellulose molecules (Keegstra et al., 1973; Valent and Albersheim, 1974; Levy et al., 1997). Binding strength between cellulose and hemicellulose varies considerably (Jarvis, 2011). The range of binding strength is likely the result of variation in the degree to which hydrogen bonding occurs along the length of a cellulose molecule and its hemicellulose binding partner. Most hemicellulose chains are not bound to cellulose along the full length of the molecules but instead are commonly tethered along segments in fibril–fibril interactions (Bootten et al., 2004). For example, the repeating unit structure of XG polysaccharides facilitates the tethering of XG molecules at intervals along the cellulose microfibrils promoting an orderly cellulose network (Levy et al., 1991). Although cellulose–hemicellulose covalent links are less prevalent than association via hydrogen bonding, strength in the matrix is significantly increased by cross-linking these two core components via third party interactions—specifically, lignin or pectin cross-linkages.

1.2.6   Cross-linking: Lignin

Lignin molecules can associate with other lignin molecules forming three-dimensional cross-linked structures. Cell wall-associated enzymes including peroxidases and laccases promote oxidative cross-linking between monolignols (Fagerstedt et al., 2010; Berthet et al., 2011). Although several types of lignin–lignin linkages can occur, β-*O*-4 aryl ether linkages tend to be prevalent in such structures (Pearl, 1967; Adler, 1977). The self-association of lignin molecules to form such networks provides strength and stiffness to plant tissues and fibers (Voelker et al., 2011). Notably, lignin also links to other cell polymers, which significantly enhances cell wall rigidity. Lignin forms associations with hemicellulose, cellulose, pectin, and other cell wall molecules (e.g., protein) via a complex system of covalent cross-linking. General types of cross-linking include ether and ester bond between lignin and polysaccharides. Direct ester linkages often occur between hydroxyl groups on lignin surfaces and uronic acids such as glucuronoxylans or rhamnogalacturonans. As cross-linking

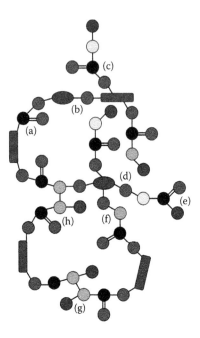

Figure 1.3 Schematic diagram showing possible covalent cross-links between polysaccharides and lignin in walls. O, PCA; •, FA; •—•, dehydrodiferulic acid. (a) Direct ester-linkage, (b) direct ether-linkage, (c) hydroxycinnamic acid esterified to polysaccharides, (d) hydroxycinnainic acid esterified to lignin, (e) hydroxycinnamic acid etherified to lignin, (f) FA ester-ether bridge, (g) dehydrodiferulic acid diester bridge, and (h) dehydrodiferulic acid diester-ether bridge. (Adapted from Iiyama, K. et al., *Plant Physiol.*, 104, 315–320, 1994.)

between cellulose and hemicellulose is generally noncovalent in nature, lignin cross-linking is a key to cell wall stability (Iiyama et al., 1994). There are three main mechanisms by which lignin is covalently linked to hemicellulose in plants (Figure 1.3). First, ester–ether bridges involving ferulic acid cross-link lignin and hemicellulose molecules (Scalbert et al., 1985; Lam et al., 1992a).

Ferulic acid, a hydroxycinnamic acid produced via the phenylpropanoid synthesis pathway, will link to lignin via ether bonds while simultaneously forming ester linkages to sugar residues within hemicellulose (Gubler et al., 1985; Iiyama et al., 1994; Sun et al., 2002). Its esters attach to residues of xylans, such as arabinofuranosyl or xylopyranosyl residues (Ishii and Hiroi, 1990; Levigne et al., 2004). Esters of ferulic acid may form dimeric complexes that cross-link arabinoxylan chains. It has been suggested that dehydrodiferuloyl dimers containing diester linkages between polysaccharides can cross-link to lignin via ether bonds (Hartley et al., 1990; Lam et al., 1992b). Other esters, linked with lignin, are present in cell wall including esters of *p*-coumaric acid; however, these do not cross-link lignin and hemicellulose as ferulic acid (FA) does (Lam et al., 1992a, 1994; Sun et al., 2002). Similar to ferulic acid, *p*-coumaric acid is also a hydroxycinnamic acid synthesized via the phenylpropanoid biosynthetic pathway. A second way lignin associates with hemicellulose is via lignin alcohol ether bonds with OH groups of polysaccharides (Watanabe et al., 1989). For example, a study using woody material from gymnosperms and angiosperms showed that glucose and mannose residues form ether linkages with lignin alcohols (Lam et al., 1990). The third way in which lignin interacts with hemicellulose in the cell wall structure involves uronyl ester bonds between the hydroxyl groups of lignin alcohols and uronic acid on 4-*O*-methyl-D-glucuronic acid residues in hemicellulose (Das et al., 1984; Watanabe and Koshijima, 1988).

1.2.7 Cross-linking: Pectin

Plant cell wall properties are also enhanced by the cross-linking of pectic polysaccharides to other cell wall components. Pectin cross-links not only provide additional structural complexity but also influence function mainly by impacting cell wall porosity (Baron-Epel et al., 1988; Willats et al., 2001) and plant morphogenesis (Derksen et al., 2011). Pectin polysaccharides are cross-linked through two different mechanisms. First, a noncovalent calcium–pectin cross-link may arise from an association of Ca^{2+} cations with a negatively charged pocket formed between the carboxyl groups of two galacturonic acid residues in HG chain lacking ester groups (Braccini et al., 1999; Willats et al., 2001). Runs of at least seven unesterfied galacturonic acid residues are required for calcium–pectin cross-linking (Powell et al., 1982). Second, covalent cross-linking via ester linkages with phenolic compounds is a common pectin cross-linking mechanism (Zaidel and Meyer, 2012). For example, ferulate and *p*-coumarate cross-link to pectic polysaccharides via ester bonds in spinach cell wall structures (Fry, 1979, 1983). Furthermore, borate diester bonds are involved in the dimerization of RG-II molecules (Zaidel and Meyer, 2012). Apiofuranosyl residues on side chains of RG-II participate in this cross-linking (Ishii et al., 1999). It has also been suggested that transesterification reactions mediated by methyltransferases can form uronyl ester linkages between methyl-esterified HG chains (acting as donors) and other HG molecules (Zaidel and Meyer, 2012).

In addition to intraspecies linkages, it has been further proposed that covalent cross-linking is common between the four main types of pectic polysaccharides (Figure 1.4)—HG, RG-I, RG-II, and/or XGA

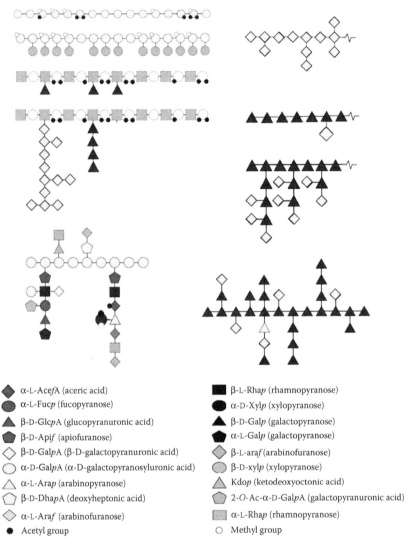

Figure 1.4 Schematic representative structures of the constituent polysaccharides of pectin. (a) HG, (b) XGA, (c) RG-II, (d) RG-I, (e) AGI, (f) arabinan, and (g) AGII. The predominant linkage types are indicated in the text. (Adapted from Vincken, J.P. et al., *Plant Physiol.*, 132, 1781–1789, 2003.)

(Yapo, 2011). The traditional model for such interspecies linkages suggests that pectin structure is composed of alternate smooth regions (HGs) and hairy regions (branched RG-Is and XGA, if present) (Aspinall, 1969; Talmadge et al., 1973; De Vries et al., 1982; Prade et al., 1999; Coenen et al., 2007). It has also been suggested that RG-I serves as a scaffold to which pectic polysaccharides such as HG, which contains RG-II elements, and XGA may attach as side chains (Vincken et al., 2003). In addition, a recent hypothetical model has been proposed in which two unbranched HG blocks and one RG-I core connected to one another could act as a scaffold. The latter is decorated with side chains such as XGA (Yapo, 2011). Ample evidence indicates that pectic polysaccharides can cross-link to nonpectin molecules in the cell wall. For example, it has been shown that pectin cross-links with hemicellulose. Specifically, HG can cross-link with XG molecules via glycosidic bonds (Thompson and Fry, 2000; Marcus et al., 2008). It has also been shown that the neutral side chains of RG-I (i.e., arabinan) can interact with the XG network (Fu and Mort, 1997; Popper and Fry, 2005, 2008). Pectin–XG complex formation facilitates retention of XG in the cell wall (Popper and Fry, 2008). HG and RG-I neutral side chains also interact with cellulose presumably via hydrogen bonding (Selvendran and Ryden, 1990; Iwai et al., 2001; Zykwinska et al., 2005; Wang et al., 2012c).

Pectin cross-linking between pectic polysaccharides; different pectic polysaccharides; and between pectin and molecules such as hemicellulose, cellulose, lignin, and protein illustrates the complex molecular networks and intermolecular associations within the cell walls of plants. It also provides a map to aid in the development of strategies for degrading plant biomass. Hydrolyzing pectin is desirable because it impacts accessibility to other cell wall components, which are targets for enzymatic degradation. The properties of plant cell walls, which are often described as a cellulose–hemicellulose network embedded in a pectin (or lignin) matrix, suggest that pectins may mask cellulose and/or hemicellulose, preventing degradative enzyme access (Varner and Lin, 1989; Cosgrove, 2001; Marcus et al., 2008; Dick-Perez et al., 2011).

Equally important are the sugars contained within the pectin itself, which represent a secondary source of fermentable sugar. Therefore, to efficiently produce biofuels from pectin-rich feedstocks, optimization of methods for pectin extraction and degradation is also necessary. Similar to starch, pectins are largely water soluble and are relatively easy to degrade when compared to other wall components. As pectins are abundant in waste residues of fruits and vegetables, pectin-rich biomass is a viable alternative feedstock for ethanol production. In order to employ pectin-rich materials as bioenergy feedstock, saccharification and fermentation methods that are optimized for the type of sugars contained within the feedstock will be required. Efforts are already ongoing to develop microbial bioprocessing strains and biomass deconstruction strategies that are specifically suited for the use of these materials (Edwards et al., 2011). Pectin-rich biomass such as citrus waste (Angel Siles Lopez et al., 2010; Pourbafrani et al., 2010), sugar beet pulp (Rorick et al., 2011), and apple pomace (Canteri-Schemin et al., 2005) has been analyzed as potential bioenergy feedstock.

1.3   Summary: Molecular substrates

The development of suitable feedstock for alternative liquid fuel production is heavily dependent on the molecular composition of the biomass from which convertible substrates will be extracted. Whether considering feedstock options for biodiesel or bioethanol, the percent composition of convertible substrates is a principal consideration in feedstock selection. Notably, the ease with which molecular substrates can be extracted from raw feedstock biomass is equally important. In corn- and sugarcane-based ethanol production, percent composition of starch and sucrose, respectively, provides an indication of the amount of energy that can be generated per unit weight of raw biomass. However, for lignocellulosic-based ethanol, the percent composition of cellulose, hemicellulose, lignin, and pectin is only a first step in the feedstock selection process. The nature of the lignocellulosic matrix is also a critical factor. The percent composition of convertible substrate (e.g., cellulose) provides only a rough estimate of net energy yield. Ultimately, it is the composition *and* structure of the matrix that determines the effective energy yield per unit weight of raw biomass and thus the economic viability of the candidate feedstock.

This is due to the fact that the structure impacts accessibility to convertible substrates (Figure 1.5) and determines the cost of extracting key molecular components (e.g., cellulose and hemicellulose), which are then used to generate final end-product substrates (e.g., glucose) through a series of chemical processes and enzymatic reactions. For first-generation biofuel (FGB) feedstock, this is less of a concern because starch and sucrose are readily extracted and converted. For second-generation biofuel (SGB) feedstock, the composition and structure of lignocellulose are often the determining factors in determining whether a feedstock is viable. Furthermore, cellulosic biomass contains secondary components including organic compounds such as terpenes, fatty acids, waxes, phenols, tannins, and flavonoids and inorganic compounds, many of which turn to ash upon combustion (Vassilev et al., 2012). Despite the apparent complexity involved in developing SGB versus FGB feedstock, the abundance of lignocellulosic biomass and the ability to avoid the *food-versus-fuel* dilemma justify efforts to develop viable SGB feedstock.

In Chapter 2, both FGB and SGB feedstock are considered in terms of abundance (i.e., global availability) and composition. For SGB feedstock, percent composition for key molecules—cellulose, hemicellulose, pectin, and lignin—will be addressed. Whereas cellulose, hemicellulose, and pectin can be converted to fermentable substrates (i.e., simple sugar), lignin drives the recalcitrant nature of cellulosic materials impacting access to convertible substrates.

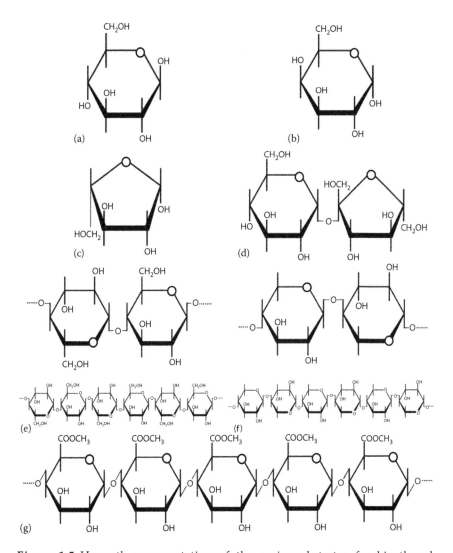

Figure 1.5 Haworth representation of the main substrates for bioethanol. (a) β-D-glucose, (b) α-D-glucose, (c) fructose, (d) sucrose, (e) cellulose, (f) xylan, (g) homogalacturonan.   (*Continued*)

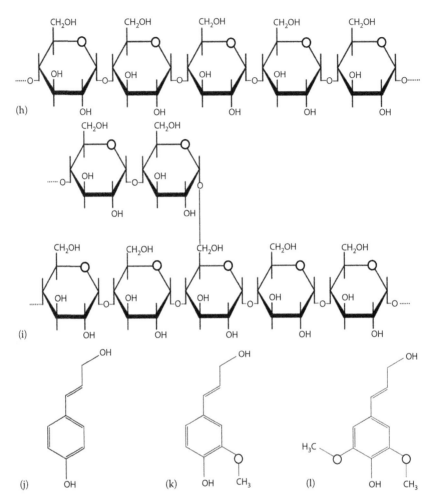

Figure 1.5 (Continued) Haworth representation of the main substrates for bioethanol. (h) amylose, (i) amylopectin, (j) *p*-coumaryl alcohol or *p*-hydroxyphenyl alcohol (monolignol), (k) coniferyl alcohol (monolignol), and (l) sinapyl alcohol (monolignol).

chapter two

First-generation biofuel and second-generation biofuel feedstocks
Biofuel potential and processing

Molecular substrates for first-generation biofuel (FGB) and second-generation biofuel (SGB) are derived from natural feedstocks. All biofuel feedstocks require some degree of processing to extract core substrates. Both the extraction efficiency and the ease (or difficulty) with which such substrates are decomposed to fermentable sugars are, in part, what determines the commercial competitiveness of a particular feedstock. Agricultural inputs, land utilization, and trade-offs between food crop versus fuel crop production are also determinants of feedstock commercial competitiveness (Sanchez and Cardona, 2008). Ultimately, the potential of any feedstock is tested by the market. Therefore, maximizing substrate extraction and reduction while minimizing processing steps and their associated energy inputs is the key to developing a viable alternative solution for meeting global liquid fuel demands. Each of the major FGB and SGB feedstocks has unique advantages (and disadvantages) in terms of processing.

2.1 Corn and sugarcane

Presently, bioethanol is produced almost exclusively from FGB monoculture crops including corn, sugarcane, sugar beet, and sweet sorghum (International Energy Agency, 2011; Nigam and Singh, 2011). In 2013, about 85% of the global bioethanol supply is produced by the United States and Brazil with corn (*Zea mays* L.) and sugarcane (*Saccharum* L.) serving as the primary feedstocks, respectively (Renewable Fuels Association, 2014). Corn, more appropriately called maize, comes in many varieties with similar composition (Table 2.1).

Table 2.1 Composition of common maize types (% composition)

Maize type	Moisture	Ash	Protein	Crude fiber	Ether extract	Carbohydrate
Salpor	12.2	1.2	5.8	0.8	4.1	75.9
Crystalline	10.5	1.7	10.3	2.2	5.0	70.3
Floury	9.6	1.7	10.7	2.2	5.4	70.4
Starchy	11.2	2.9	9.1	1.8	22	72.8
Sweet	9.5	1.5	12.9	2.9	3.9	69.3
Pop	10.4	1.7	13.7	2.5	5.7	66.0
Black	12.3	1.2	5.2	1.0	4.4	75.9

Source: Food & Agriculture Organization of the United Nations, Gross chemical composition, in *Maize in Human Nutrition*, David Lubin Memorial Library, Rome, Italy, 25, 1992.

2.1.1   Corn

Corn grain is one of the two main feedstocks for which infrastructure and production technology exists to effectively extract substrate (i.e., starch) to produce marketable (i.e., cost-competitive) bioethanol. There are approximately 200 major bioethanol plants in the United States that produce an estimated 13.3 billion gal (bg) of corn-based ethanol in 2013 (Renewable Fuels Association, 2014). Corn-based ethanol accounts for about 10% of the nation's transportation fuel needs (Hill et al., 2006). In spite of the success of the corn-based ethanol industry, ethanol cannot be currently produced neither at a low enough cost per gallon nor in sufficient quantities to serve as a complete substitute for traditional fossil fuel-based products. It has been argued that this is partially a consequence of the fact that corn-based *fuel* products compete with corn-based *food* products. Whether for fuel or for food, fertile soils are required to produce corn. This results in so-called the *food-versus-fuel* dilemma (Runge and Senauer, 2007). Large-scale commitments to corn ethanol production require either increasing the farm land development, which impacts land use and biodiversity, or diverting the harvested grain to ethanol production (rather than food production), which can cause food shortages and fluctuation in food prices (Landis et al., 2008; McDonald et al., 2009; Reijnders and Huijbregts, 2009; Sala et al., 2009).

The impact of higher corn prices on corn-based food products for direct human consumption (e.g., breakfast cereal) may be modest; however, indirect effects such as an increase in the price of corn-based livestock feed, which, in turn, influence the prices of meat and dairy products, result in a significant net impact. Although estimates vary, food prices could readily increase by 1%–10% with a significant increase in demand for corn ethanol (Perrin, 2008).

Despite the *food-versus-fuel* dilemma, the land and water requirements, the agricultural inputs (e.g., fertilizers), and the fact that ethanol greenhouse gas (GHG) emissions do not appear to be significantly less than GHG emissions from gasoline combustion, the corn ethanol industry appears to have carved out a niche in the trillion dollar per year alternative fuel market that is sustainable over the long term. Moreover, there is a potential to further reduce the per-gallon production costs with new and emerging technologies.

2.1.2   Sugarcane

The second most lucrative feedstock in bioethanol production is sugarcane, and Brazil represents the largest producer of sugarcane-based ethanol with ~79% derived from sugarcane juice and the remaining volume produced from residual molasses (Wilkie et al., 2000; Food and Agriculture Organization of the United Nations, 2008). Ethanol production from sugarcane proceeds by cleaning raw feedstock, mechanically extracting the sugars, concentrating the juice, fermentation, distillation, and dehydration (Dias et al., 2011). Extraction of sugarcane juice, which contains sucrose and other simple sugars (Table 2.2), requires only mechanical pressing with no need for high temperature, chemical, or enzyme pretreatments. With such advantages, it would seem that sucrose is the ideal molecular substrate, and sugarcane is the ideal feedstock for ethanol production. However, there are several important factors that limit sugarcane-based ethanol production.

First, sugar from sugarcane is a global commodity. Therefore, the diversion of sugar production to fuel production is carefully monitored as a component of the *food-versus-fuel* dilemma. Producers can strike a balance by extracting sugar and then producing ethanol from molasses. Indeed, many ethanol plants in Brazil produce ethanol from molasses A, which is the residual product following a single crystallization step for sugar removal (Ensinas et al., 2009). Fermentation of molasses A typically yields a ~9% v/v ethanol concentration. As the molasses is naturally low in free nitrogen treatments such as urea supplementation, it is incorporated

Table 2.2 Chemical composition of sugarcane (% composition)

Major components	Percent dry wt.	Minor components	Percent dry wt.
Water	71.57	Aconitic acid	1.79
Sucrose	13.30	Glucose	0.62
Cellulose	4.77	Dirt	0.60
Hemicellulose	4.53	K_2O	0.20
Lignin	2.62		

Source: Furlan, F.F. et al., *Comput. Chem. Eng.*, 43, 1–9, 2012.

into the process to ensure that yeast-based fermentation is efficient. Other nutrients such as phosphorus, biotin, pantothenic acid, and inositol may also be added (Piggot, 2003). Although employing an initial crystallization step allows the production of both sugar for food and molasses for fuel, this is not a solution for *food-versus-fuel* because subsequent crystallization from molasses A generates additional sugar.

Second, high-yield sugarcane is most efficiently grown in tropical and subtropical regions. Large-scale sugarcane cultivation for ethanol production that meets world liquid fuel demands would require the clearing and development of new plantations in countries such as Brazil. Clearing hectares of rain forest to cultivate sugarcane reduces biodiversity and may contribute to global warming or other ecological detriment (Reijnders and Huijbregts, 2009; Delucchi, 2010). Again, as with any FGB feedstock, the benefits of producing sugarcane for ethanol must be considered holistically with factors such as freshwater use, land requirements, and ecological impacts (Delucchi, 2010).

2.2   Lignocellulosic biomass

In terms of contribution to current ethanol supplies, the least developed feedstock class is lignocellulosic biomass. Still, lignocellulosic feedstocks are arguably the most promising resource for commercial ethanol production primarily due to abundance and renewability with annual global productivity estimated between 10 and 50 billion tons per year (Goldstein, 1981; Lutzen et al., 1983). Collectively referred to as second-generation feedstocks, lignocellulosic materials may be obtained from multiple sources including energy crops, agricultural residues, forestry wastes, and municipal solid wastes (MSW; Lynd et al., 1991).

As described in Section 2.2, cellulosic biomass typically consists of cellulose, hemicellulose, lignin, pectin, and trace amounts of starch. Percent composition of each component varies depending upon the feedstock type (Table 2.3). Unlike corn and sugarcane ethanol, cellulosic ethanol is currently not commercially competitive. Despite the availability of numerous feedstocks (most of which do not compete with food crops), less than 10% of commercially available bioethanol is produced from lignocellulosic materials. The recalcitrant structure, or *matrix*, formed by the main components of lignocellulosic biomass (i.e., cellulose, hemicellulose, and lignin) inhibits the transformation of these materials to readily fermentable sugar (Himmel et al., 2007). Present technology requires the use of physical and chemical pretreatments as well as enzymatic processes to break down these complex lignocellulosic structures (Chundawat et al., 2011). Economic viability of cellulosic ethanol at an industrial scale has been hindered by the costs of such pretreatments and enzymatic processes, which are required to access key polysaccharides and convert substrate to a refined ethanol product (Wooley et al., 1999; Aden et al., 2002; Yang and Wyman, 2008).

Table 2.3 Composition of selected cellulosic feedstocks

Feedstock	Cellulose (% dry wt.)	Hemicellulose (% dry wt.)	Lignin (% dry wt.)
Bamboo[a]	49–50	18–20	23
Corncob[b]	35	35–42	5–15
Corn stover[b]	39–42	19–25	15–18
Cotton seed hairs, flax[b]	85–95	5–20	0
Cotton stalk[a]	31	11	30
Coffee pulp[a]	33.7–36.9	44.2–47.5	15.6–19.1
Douglas fir[a]	35–48	20–22	15–21
Eucalyptus[a]	45–51	11–18	29
Hardwood[b]	45–47	25–40	20–55
Rice straw[b]	32–47	15–27	5–24
Rice hulls[b]	24–36	12–19	11–19
Wheat straw[b]	30–49	20–50	8–20
Wheat bran[a]	10.5–14.8	35.5–39.2	8.3–12.5
Grasses[a]	25–40	25–50	10–30
Newspaper[b]	40–55	25–40	18–20
Sugarcane bagasse[a,b]	25–45	24–32	12–25
Sugarcane tops[a]	35	32	14
Pine[a]	42–49	13–25	23–29
Poplar[c]	42–49	16–23	21–29
Olive trees biomass[a]	25.2	15.8	19.1
Jute fibers[a]	45–53	18–21	21–26
Switchgrass[b]	30–50	10–40	5–20
Grasses[b]	25–40	25–50	10–30
Rye straw[b]	30.9	21.5	25.3
Oilseed rape[a]	27.3	20.5	14.2
Softwood[b]	40–45	25–29	30–60
Oat straw[a]	31–35	20–26	10–15
Nut shells[a]	25–30	22–28	30–40
Sorghum straw[a]	32–35	24–27	15–21
Tamarind kernel powder[a]	10–15	55–65	–
Water hyacinth[a]	18.2–22.1	48.7–50.1	3.5–5.4

[a] Menon, V. and Rao, M., *Prog. Energy Combust. Sci.*, 38, 522–550, 2012.
[b] Paulová, L. et al., Production of 2nd generation of liquid biofuels, in *Liquid, Gaseous and Solid Biofuels—Conversion Techniques*, Z. Fang (Ed.), InTech, 2013.
[c] Sannigrahi, P. et al., *Biofuels Bioprod. Biorefin.*, 4, 209–226, 2010.

Furthermore, processes that employ ethanologenic microorganisms for degrading cellulosic materials (and releasing sugars) are not as efficient as the microbial systems (e.g., yeast systems) used for conversion in corn- and sugarcane-based processes, because most conventional ethanologenic microorganisms are not able to fermenting all the sugars released during the processing of biomass into ethanol (Barnett, 1976). Nonetheless, the potential of large-scale production of cellulosic ethanol is generally recognized. Price spikes in petroleum-based liquid fuels, increased demand for corn- or sugarcane-based ethanol, and emerging technologies that overcome current production challenges position the cellulosic ethanol as the preeminent liquid fuel product of the future.

In addition to benefiting from a diverse selection of abundantly available high-yield feedstocks, cellulosic ethanol offers significant reductions in GHG emissions (Michael et al., 2012). Lignocellulosic ethanol generates 91% less GHG than fossil-based petrol or diesel in transport applications. As a comparison, reduction in GHG emissions for corn ethanol compared to fossil fuel is 22% (United States Environment Protection Agency, 2007). Although cellulosic feedstocks still require land and water resources, there is no direct *food-versus-fuel* dilemma with cellulosic ethanol production since the feedstocks are not generally consumed as food. Thus, there are social and environmental benefits to cellulosic ethanol production that are not realized with either petroleum-based or corn- and sugarcane-based liquid fuel products.

The *carbon footprint* of cellulosic ethanol production is significantly lower than petroleum-based fuels and corn-based ethanol. Cellulosic ethanol has the additional advantage that most of the by-products generated during cellulosic ethanol production can also be utilized. A wide spectrum of chemicals and materials can be generated from lignocellulosic biomass. Effective planning and design (or retooling) of cellulosic-based refineries could provide a sustainable source of commercializable organic compounds, synthetic polymers, pharmaceuticals, household products, and other coproducts generated from cellulosic ethanol production (Patton, 2010).

Alternatively, lignin-rich residues generated from cellulosic ethanol production can be burned to produce electricity. This electricity can either be used for power production or be diverted back to the power grid and sold (Kim and Dale, 2004). From an economic standpoint, it has been estimated that the development of bioenergy production technologies could result in 9.7 million new jobs by 2030 (International Renewable Energy Agency, 2013). Many of those jobs could either be directly or indirectly associated with cellulosic ethanol production.

Thus, the question begs: If there is an abundance of high-yield cellulosic materials, a reduction in GHG emissions, a reduced carbon footprint, a reduced competition with food crops, the potential for producing valuable coproducts, the potential for new employment opportunities and thus general economic and social benefit, then why is cellulosic ethanol *not* a major fuel commodity? As mentioned earlier, the answer to this question lies in the recalcitrant nature of the ligno-cellulosic matrix. Ultimately, the key substrates for cellulosic ethanol production are cellulose and hemicellulose. These long chain carbo-hydrates must be hydrolyzed to produce simpler fermentable sugars. Although hemicellulose has a lower molecular weight than cellulose, its short lateral side chains comprised of various sugar moieties result in a more complex degradation profile than cellulose. As a simple poly-mer of glucose units is organized in oligomeric subunits within the chain, linear chains of cellulose are readily hydrolyzed releasing glu-cose oligomers and monomers. However, accessing cellulose within the lignocellulosic matrix of various feedstocks has proven to be a formidable challenge for those who aspire to produce competitively priced cellulosic ethanol products. Specifically, enzymatic hydrolysis is inhibited by the complex entanglement and dense structure resulting from covalent cross-linking between cellulose, hemicellulose, lignin, pectin, and other molecular constituents within the matrix. Indeed, many lignocellulosic structures are often compared to a fiberglass-resin matrix in which long chains of cellulose are *shielded* by strands of hemicellulose and are bound in lignin *glue*. In nature, this association provides the mechanical strength and physical protection that plants require for proper growth and resistance to predators and pathogens. Moreover, each feedstock is unique in both its content and composi-tion of these molecules, making a single solution for use of multiple feedstocks in a single production process near impossible (with cur-rent technology). In short, the association of hemicellulose, lignin, and cellulose produces sturdy plant cell walls—an evolutionary benefit for living plants and a problem leading to higher production costs for cel-lulosic ethanol producers (Zhang, 2008). Still, there are some cellulosic ethanol products that are making it to market, and new technologies are emerging to deal with feedstock breakdown issues. Some feed-stocks have proven particularly promising. These include corn stover, bagasse, crop straw, grasses, woody energy crops, forestry wastes, municipal wastes, as well as several other cellulosic materials, which are abundant and that can be deconstructed to yield readily ferment-able sugar. As biofuel technology progresses, several such feedstocks could play a central role in supplementing liquid fuel demands and in offsetting some of the deleterious impacts of reliance on fossil fuels.

2.2.1   Corn stover

Corn stover is an abundant waste product comprised of stems, leaves, and cobs resulting from corn production. In the United States, corn sto-ver is one of the most abundant agricultural waste products generated at approximately 196 million tons available annually (Graham et al., 2007). Dry corn stover can be as much as ~40% cellulose by weight with relatively low lignin content. Hemicellulose content in corn stover can be as high as ~25% by dry weight (Paulová et al., 2013). Rich in cellulose and hemicellulose and relatively low in lignin, corn stover is a prime feedstock for cellulosic ethanol. In spite of a *relatively* low lignin content (when compared to other feedstock options), per pound of dry weight, lignin in stover is present at a comparable percent composition to hemicellulose. Although hydrogen bonding between cellulose and hemicellulose is disrupted with relative ease, cross-linking of lignin to these polysaccharides renders corn stover recalcitrant to sugar reduction. In other words, the lignin content results in reduced access to cellulose and hemicellulose, the two key substrates for producing readily fermentable sugar. Still, given its vast availability (especially in the Midwestern United States), corn stover is considered one of the more economically promising cellulosic feedstocks, and significant research has been focused on deconstructing its lignocellulosic matrix. All of the more effective protocols require some type of pretreatment of the stover to disrupt the matrix prior to enzymatic treatments, which are in turn designed to release and reduce cellulose (and hemicellulose) from the matrix (Section 2.3). With 196 million tons (~400 billion lb) of corn stover generated per year in the United States and with a theoretical yield of ~113 gal of ethanol per ton of stover (~18 lb corn stover/gal ethanol), as much as 21.8 billion gal of stover-based ethanol could be produced per year.

2.2.2 *Bagasse*

Bagasse is the residual fibrous material that remains after juice is extracted from sugarcane stalk. Bagasse is rich in cellulose, hemicellulose, and lignin. Cellulose content of bagasse varies widely but typically ranges from 25% to 45% of total dry weight (Menon and Rao, 2012; Paulová et al., 2013). Percent composition of hemicellulose in bagasse is similar to what is found in corn stover. However, lignin content in bagasse is higher (on average) than what is present in corn stover. The moderately high lignin content in bagasse is the factor that makes it a challenging feedstock for bioethanol production. As bagasse has no food value, it does not directly contribute to the *food-versus-fuel* dilemma and may be used as a high-yield feedstock for ethanol production. However, current industrial trends exploit the availability of bagasse to produce heat and steam via combustion for evaporating water during the crystallization process of sugar production and to distill the ethanol (Siddhartha Bhatt and Rajkumar, 2001).

After juicing, ~0.3 lb of wet bagasse is produced (with approximately 50% moisture content) from each pound of wet sugarcane (Shapouri et al.,

2006). With the annual global production of sugarcane at 4 trillion lb (estimated from sugar production at 175 million metric tons per year), over 1.2 trillion lb of bagasse is available annually (USDA Foreign Agricultural Service, 2014). (Note: In a pound of sugarcane, ~85% is juice by weight, and 11% of the juice is sugar by weight). It takes approximately 18 lb of raw bagasse to produce a gallon of bagasse-based bioethanol. Therefore, ~70 billion gal per year of ethanol could be produced if all bagasse was used to produce ethanol. Due to high abundance, bagasse could serve as another primary alternative source for liquid fuel in the future.

2.2.3   Crop straw

Upon removing the grain (and chaff) of cereal plants, a residual stalk or *straw* remains. Although straw is a by-product of cereal crop production and stalk biomass varies depending upon the specific crop species harvested, on average ~1.5 lb of straw is generated per pound of grain (1.3 kg straw/kg grain) (Peterson, 1988). Yield not only varies depending upon species but is also impacted by agronomic and climatic factors. With regard to chemical composition, straw predominantly consists of cellulose (32%–47%), hemicellulose (19%–27%), and lignin (5%–24%) (Maiorella, 1983; Zamora and Sanchez Crispin, 1995; Garrote et al., 2002; Saha, 2003a). Pentoses dominate the hemicellulose with xylose being the most abundant sugar (14.8%–20.2%) (Maiorella, 1985; Roberto et al., 2003). The most abundant sugars are rice and wheat straw. Annual production of rice straw across Africa, Asia, Europe, and America is ~731 million tons (Kim and Dale, 2004). At least an equivalent amount of straw is generated for each pound of rice grain harvested, and in some cases straw mass can exceed grain by a factor of 1.5 (Maiorella, 1985). In terms of biomass-to-ethanol yield, rice straw is comparable to stover and bagasse; in that, approximately 18 lb of rice straw is required to produce a single gallon of bioethanol. Considering an annual global production of 731 million tons, (~1.5 trillion lb), it is estimated that ~55 billion gal/year of ethanol could be produced using rice straw (Kim and Dale, 2004). Rice straw has several characteristics that make it a high potential ethanol feedstock including a high cellulose and hemicellulose content with relatively low total alkali content (Baxter et al., 1996; Paulová et al., 2013). The major polysaccharides are readily hydrolyzed into fermentable sugars, especially in strains that exhibit low lignin content. In strains with high lignin content and the presence of high ash and silica content (~15% and 75%, respectively) do present challenges in using rice straw for bioethanol production (Zevenhoven, 2000). However, as a highly abundant renewable resource, it remains a viable feedstock option (Kim and Dale, 2004).

Second only to rice straw, wheat straw is also an abundant and viable feedstock for bioethanol. In Europe and United States, wheat straw is the most abundant straw-based agricultural waste product. Wheat straw generally contains less ash on an average (~9%) than rice straw (~12%) (Lee, 1997; Peiji et al., 1997; Karimia et al., 2006).

However, the ash tends to contain high levels of silica (up to 55%) and alkali content (>25%), which must be considered during feedstock deconstruction and substrate conversion processes (Baxter et al., 1996; Zevenhoven, 2000). Feedstock with high alkali content is less susceptible to enzymatic degradation, a key process in deconstructing cellulosics (Dhillon and Khanna, 2000). Wheat straw is generally of 30%–49% cellulose, 20%–50% hemicellulose, and 8%–20% lignin by mass (Paulová et al., 2013). Depending on the processing methods, ethanol yield from wheat straw may be as high as 84%. Beyond feedstock pretreatment, other factors can influence ethanol yield including enzyme selection and loading, growth conditions (which impact biomass composition), and the yeast used (Detroy et al., 1982; Delgenes et al., 1990; Nigam, 2001; Saha and Cotta, 2006; Kim et al., 2009b). At least an equivalent amount of straw is generated for each pound of wheat grain harvested. In some cases, straw mass can exceed grain by a factor of 1.3 (Peterson, 1988). In the United States alone, ~86 million tons (~172 billion lb) of wheat straw are produced annually. Adding production from other parts of the globe, total annual wheat straw production is approximately ~744 million tons/year (~1.5 trillion lb). It is estimated that ~30 lb of wheat straw is required to produce a single gallon of ethanol. Wheat straw-based ethanol could top 50 billion gal per year. Together Asia and Europe could potentially provide over 60% of the worldwide straw-based ethanol supply (Table 2.4).

Table 2.4 Regional potential bioethanol production from rice straw and wheat straw

Potential bioethanol production	Rice straw GL/y (bgy)[a]	Wheat straw GL/yr (bgy)[a]
Africa	5.86 (1,548)	1.57 (415)
Asia	186.8 (49,347)	42.6 (11,254)
Europe	1.10 (291)	38.9 (10,276)
North America	3.06 (808)	14.7 (3,883)
Central America	0.77 (203)	0.82 (217)
Oceania	0.47 (124)	2.51 (663)
South America	6.58 (1,738)	2.87 (758)
World	**204.64 (54,060)**	**103.97 (27,466)**

Source: Kim, S. and Dale, B.E. *Biomass Bioenergy.*, 26, 361–375, 2004.

[a] GL/yr, gigaliters per year; bgy, billion gallons per year (rounded).

2.2.4 Grasses (perennial energy crops)

Perennial energy crops (PECs) are low-cost/low-maintenance crops (typically grasses) that are used as cellulosic feedstock for ethanol production. As alternative to *low net carbon* feedstock, high-yield PECs may be grown on marginal lands that are not suitable for growing food crops. With reduced water requirements (when compared to food crops) and minimal maintenance, PECs can be grown with minimal environmental impact while avoiding food-versus-fuel issues (Schmer et al., 2008; Dominguez-Faus et al., 2009; Cai et al., 2011). Herbaceous PECs represent a viable source of carbohydrates for bioethanol (Dien et al., 2005, 2008). Indeed, lignocellulose from perennial species, especially C4 grasses, is possibly better suited over grain or sugar from annuals for renewable energy production in temperate zones (Carruthers et al., 1991; Cherney et al., 1991). The economic viability of PECs is supported by the fact that when compared to grain crops, there are lower production costs and energy inputs (e.g., less fertilizer and pesticide) as well as there is an improvement of soil quality of marginal crop lands (Cherney et al., 1991; Vadas et al., 2008). Species of the genus *Miscanthus* and switchgrass (*Panicum virgatum* L.) are two C4 plants that have been extensively studied as potential energy crops for European and American biofuel markets (Lewandowski et al., 2000; Schmer et al., 2008). *Miscanthus* spp. are rhizomatous perennial grasses, which have been shown to be exceptionally tolerant to cold climates (Beale and Long, 1995; Beale et al., 1996; Naidu and Long, 2004; Spence et al., 2014). Evolutionarily, it is proposed that *Miscanthus* is originated in tropical regions of East or Southeast Asia. However, it is found globally and grows across a broad geographic and climatic range extending from the Pacific Islands to the mountainous regions of Japan (Singh et al., 1983; Clifton-Brown et al., 2008). Interestingly, it is a naturally occurring sterile triploid hybrid, *M. × giganteus*, derived from the crossing of *M. sacchariflorus* and *M. sinensis* that has attracted the attention of scientists due to its ability to grow rapidly, low maintenance requirements, and its high ethanol yield profile (Lewandowski et al., 2000; Hodkinson et al., 2002). Under experimental conditions, it has been demonstrated that this hybrid can produce 2.6 times more ethanol per unit land area, while requiring significantly fewer inputs, than corn grain (Heaton et al., 2008).

Switchgrass is a member of the *Paniceae* tribe of grasses and is a member of the family *Poaceae*. Similar to *Miscanthus*, switchgrass is also a perennial C4 grass that grows robustly on lands that are considered unsuitable for common agricultural crops. Switchgrass roots deeper than many other grasses and grows from central Mexico up into more northern climates to about 55°N latitude (Weaver, 1968; Stubbendieck et al., 1991). Its high cellulose content renders it a viable candidate crop for ethanol (and butanol) production (Paulová et al., 2013). Specifically, the fact that switchgrass is a perennial C4 grass, which can be stored either wet or dry, and that its cultivation is compatible with conventional farming methods, makes it a prime candidate as a high-yield biofuel feedstock—especially along the midsouthern eastern region of the United States (McLaughlin et al., 1999; Sokhansanj et al., 2009).

2.2.5   Bamboo

In addition to *Miscanthus* and switchgrass, a fast growing woody grass of the family *Poaceae* may soon become a major feedstock in the production of second-generation bioethanol. Bamboo, a set of large woody grasses of the subfamily *Bambusoideae*, is one of the most prolific and sustainable plants in the world. It is naturally distributed across the tropics and subtropics but also grows well in temperate climates. It is found globally (except in Europe) at latitudes from 46°N to 47°S and altitudes from 0 to 4,000 m (Soderstrom and Calderon, 1979; Williams et al., 1991; Ohrnberger, 1999). The ability of bamboo to grow in nutrient-poor soil, its minimal requirements for silviculture, the ease with which it can be harvested, the fast growth and other advantageous properties make bamboo a viable candidate as an energy crop (Huberman, 1959; Fernandez et al., 2003; NL Agency, 2013). Moreover, higher annual biomass yield and environmental benefits such as erosion control and carbon sequestration make bamboo cultivation *sustainable* and may make it one of the more economically viable feedstocks (He et al., 2014). Indeed, bamboo is often considered more sustainable than other energy crops (Guarnetti, 2007). Once established, there is no need for replanting. Harvested culms are replaced by new shoots that emerge from the underground rhizome system. This enables sustainable, regular harvesting of culms (and thus stable income for producers) with minimal investment (NL Agency, 2013). Bamboo matures within 5–7 years; however, some species can reach full height within 60 days. Rapid growth rate, low ash content, low alkali index, and high-energy content endow bamboo with key properties that are characteristic of other viable energy crops (Scurlock et al., 2000). Bamboo is already cultivated in the northeastern regions of Brazil as a dedicated energy crop and is considered to be the second (only to sugarcane) in terms of annual energy potential (30.8 TWh) (Anselmo Filho and Badr, 2004).

In bamboo, cellulose content is approximately 50%; hemicellulose ranges from 18% to 20%; and on average approximately 23% is lignin (Menon and Rao, 2012). It has also been reported that bamboo possesses 25%–30% of lignin, placing bamboo at the high end of the 11%–27% range for other nonwoody feedstock, and more closely resemble values found in the softwoods (24%–37%) and the hardwoods (17%–30%) (Bagby et al., 1971; Fengel and Wegener, 1984; Dence, 1992). Bai suggests that bamboo would be an economically viable fiber resource if lignin and its derivatives could be either removed or converted to added-value coproducts, which can be used as substitutes for selected synthetic compounds that are currently derived from oil (Bai et al., 2013). However, as with other feedstock types, the complex physical and chemical interactions between cellulose, hemicellulose, and lignin in bamboo prevent commercial enzymes from readily accessing microfibrillar cellulose. Thus, mechanical and/or chemical pretreatments are required to deconstruct the lignocellulosic matrix and to expose cell wall sugars to reducing enzymes, which are employed to hydrolyze the complex carbohydrates to reduced, readily fermentable sugars (McMillan, 1994; Himmel et al., 2007; Jørgensen et al., 2007; Zhao et al., 2012).

Although bamboo maintains a higher percent composition of hexoses compared to some other low-lignin species such as napiegrass, ethanol yields from bamboo can be comparatively low (Yasuda et al., 2013). This is likely due to poor accessibility of reducing enzymes to cellulosic components of bamboo. However, advanced pretreatment regimens (Section 2.3) can expose cellulose and hemicellulose and can increase the efficiency of deconstruction enzymes. Under such conditions, it is possible to produce up to 247 L ethanol/ dry ton of giant bamboo. Yield increases up to 292 L ethanol/dry ton of giant bamboo have been obtained when pretreatment resistant microorganisms are employed, and when maximum sugar yields (combined severity factor (CSF) = 2.25) are obtained (García-Aparicio et al., 2011).

2.2.6   *Woody energy crops and forestry waste*

Beyond straws, grasses, corn stover, and bagasse, hardwood crops and waste materials from forestry activities are also considered as a potential feedstock for commercial biofuel production. Woody energy crops (WEC) including single-stem hardwoods such as hybrid poplars, cottonwoods, eucalyptus, and sycamore trees have been studied and used for fiber and energy since the late 1960s (Dawson, 1972; Debell et al., 1972; Johnson, 1972). In 1981, the U.S. Department of Energy along with scientists at the Oak Ridge National Laboratory (ORNL) developed the Short-Rotation Woody Crops Program (SRWCP) to further investigate the potential role of WECs to fill gaps in energy needs (Ranney et al., 1987). SRWCP was the first systematic large-scale study of WECs for biofuel production undertaken in the United States. As a favorite source of pulp for the paper industry, breeding programs for poplar were initiated in the United States in the 1920s and in other developing countries in the 1920s and 1930s (Stout and Schreiner, 1933; Richardson et al., 2007). Due to favorable characteristics including growth throughout a broad range of geographic latitudes, rapid growth, drought tolerance, resistance to predators and pathogens, and chemical content (i.e., favorable cellulose and hemicellulose-to-lignin composition), poplars in particular are considered a viable WEC feedstock. As poplars have been used as feedstock for nearly a century for producing a variety of products including paper, chemicals, and adhesives, hybrids and, more recently, genetically enhanced varieties of this fast-growing short-rotation WEC have been developed (Polle et al., 2013). Cellulose content ranges between 42% and 49%; hemicellulose content ranges from 16% to 23%; and lignin ranges from 21% to 29%. Indeed, poplar has greater cellulose content than corn stover or switchgrass, making it a suitable feedstock for bioethanol production; however, it does have the undesirable property of higher lignin content (Sannigrahi et al., 2010b; Paulová et al., 2013). In terms of ash content (another undesirable factor), hybrid poplar often exhibits a higher content than the softwoods but is still substantially lower in ash when compared to corn stover, switchgrass, wheat straw, and other feedstocks (Brown, 2003).

Species of cottonwoods, aspens, balsam poplars, white poplars, and hybrids exhibit rapid growth. Many are native to Europe, North America, Asia, and northern Africa. In the United States alone, logging and other development activities generate ~176 million m^3 of residue per year (Smith et al., 2009). Although primary forest operations generate the majority of this residue via delimbing, topping, and bucking operations (Balckwelder et al., 2008), dead, rough, and rotten trees as well as small trees and non-commercial trees contribute to this total residual biomass, which is commonly not used and simply left on-site as waste. Currently, removal and use of logging residue and woody waste (from clearing and development activity) for use as energy feedstock are not considered profitable. Interestingly, successful fire suppression has resulted in forest growth and overgrowth, providing another potential source of woody feedstock because thinning operations are employed to reduce the risks of severe fires that threaten commercial areas, residential areas, and protected parks. Rising fossil fuel prices and growth in alternative liquid fuel markets may soon provide economic justification for extracting large quantities of forest residues for bioenergy production. WECs, forestry wastes, and woody waste from municipal sources together provide a substantial source of biomass for energy generation. Although it is difficult to estimate global annual yields due to uncertainties in the extent of forestry waste, woody waste from MSW, and construction and demolition waste, over the past 30 years in the United States, wood-based products were produced in an average of 143.3 million tons annually (Howard, 2012).

2.2.7   Municipal waste

Woody components of municipal waste constitute only a fraction of total MSW that is generated worldwide. MSW consists of a variety of items ranging from organic food scraps to discarded furniture, packaging materials, textiles, batteries, appliances, and other materials including yard trimmings. MSW production is particularly high in developed countries. In 2012, ~251 million tons of MSW was generated in the United States (~4.38 lb/person/day) (United States Environmental Protection Agency, 2014). Despite this significant quantity of residual biomass, the energy potential of MSW has received relatively little attention. Shi et al. (2009b) conducted an analysis of the potential to convert selected fractions of MSW to ethanol and determined that glucans and xylans may be recovered in significant quantities. Li and Khraisheh (2008) determined that selected fractions of MSW could hold up to 90% glucose per gram dry weight. When used to produce ethanol, it was demonstrated that ~60% of the theoretical recovery could be achieved (Ballesteros et al., 2010). The composition of municipal waste varies from country to country and changes significantly with time. In the United States, paper and cardboard are the largest components of MSW. Organic materials comprise the second largest component. Yard trimmings account for the third largest component (Figure 2.1). In Europe and in many countries around the globe, MSW plays a significant role for urban centers as a source of energy via waste incineration, which generally features high conversion efficiency (Themelis, 2003).

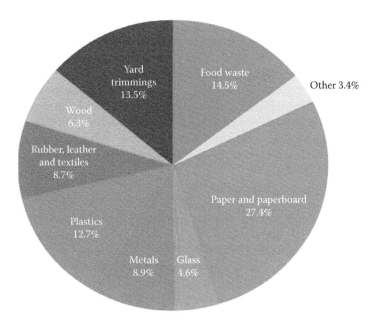

Figure 2.1 Total MSW generation (by material), 2012—251 million tons (before recycling). (Adopted from United States Environmental Protection Agency, Municipal solid waste generation, recycling, and disposal in the United States: facts and figures for 2012, 2014.)

As rural areas have low MSW density, MSW-based energy production and its use are often limited to urban settings (Tyson et al., 1996). Many sources of MSW are particularly appealing as feedstock for ethanol production because cellulosic components comprised of paper, wood, or yard waste account for 40%–70% of MSW dry weight (Li et al., 2012). As urban populations increase worldwide, management of MSW becomes more challenging because the common approaches of landfill and incineration of waste can result in adverse environmental consequences (Lisk, 1988; Lou and Nair, 2009). Effective MSW management strategies could address not only the increasing urban waste volumes but also could contribute toward diversifying feedstock sources for bioethanol production. Theoretical annual ethanol yields from waste paper alone have been estimated at ~83 billion liters (Shi et al., 2009a). Both developed and developing countries could contribute to MSW-based ethanol production (Figure 2.2). Collection systems for MSW are well established. Furthermore, fees are charged by landfills to receive waste. These *tipping* fees can range from $15 to $100 per ton (Sakamoto, 2004). Therefore, MSW as a feedstock could be available at no cost (or, even at negative cost). This makes MSW a highly cost-effective, economically attractive feedstock for ethanol production (Joshi et al., 2005). Commercialization of MSW-based bioethanol has been piloted and implemented by a variety of companies that realize the potential of this feedstock (Smith, 2013; Fiberright LLC, 2015). It has been estimated that if the waste paper component of MSW alone was diverted from landfills to cellulosic ethanol processing systems, then approximately 5.36% of global gasoline demand could be met (Shi et al., 2009a). However, for MSW to become a viable feedstock for alternative liquid fuel, management strategies would need to be streamlined to separate out usable MSW components. Furthermore, pretreatment and processing methods would need to be employed to accommodate mixed MSW biomass. Still, MSW provides another alternative feedstock that could contribute significantly to the production of large quantities of cellulosic ethanol.

2.2.8 *Cellulosic feedstock: A prospectus*

According to estimates from the U.S. Energy Information Administration, approximately 230 billion gal (5.5 billion barrels) per year of petroleum-based liquid fuel will be needed in 2014–2016 in the United States alone (Energy Information Administration, 2004). Although cellulosic ethanol via a single feedstock source is not enough to fulfill future energy demands, it is quite clear that utilization of multiple feedstocks—corn stover, bagasse, crop straw, perennial energy crops (e.g., grasses), bamboo, woody energy crops, forestry waste, MSW, and others—could indeed supply enough cellulosic ethanol to significantly reduce global dependency on fossil liquid fuel. Many of these alternative feedstock options are considered very sustainable (not just renewable). Reasons why the full

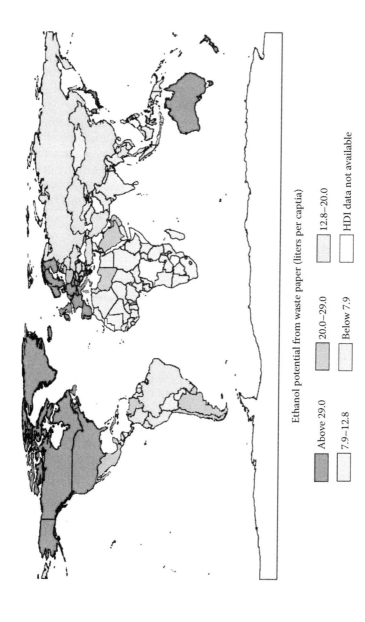

Figure 2.2 Estimated maximum cellulosic ethanol potential (liters per capita) from waste paper for 173 countries plotted onto a World map. (Adopted and modified from Shi, A.Z. et al., *GCB Bioenergy*, 1, 317–320, 2009.)

potential of cellulosic biomass has not been realized for liquid fuel production are multiple. Apart from the upfront capital required for developing new processing facilities or retooling current corn-based ethanol plants, several technical questions remain unanswered regarding the cost-effective breakdown of raw cellulosic biomass into secondary macromolecular substrates (i.e., cellulose and hemicellulose), which can in turn be hydrolyzed into readily fermentable simpler carbohydrates.

The deconstruction of raw material is a two-step process in which (1) unnecessary components of the biomass are separated from cellulosic substrates; followed by, (2) reduction of complex carbohydrates into simple, fermentable sugars. The former is often performed via physical and/or chemical feedstock pretreatment processes. The latter is typically done using enzymatic processes. It is the efficacy of these two deconstruction steps that determines whether cellulosic ethanol is economically competitive. Process efficiency and costs associated with feedstock pretreatments and enzymatic hydrolysis determine production costs and impact the price per gallon of cellulosic ethanol regardless of the feedstock being used.

2.3   Feedstock pretreatments

Numerous pretreatments have been developed with the aim of effectively releasing complex carbohydrates from raw cellulosic feedstock. Pretreatments are commonly grouped into four general categories: *chemical, physical, physicochemical,* and *biological*. Many pretreatments facilitate sugar release from raw biomass and induce the solubilization of cell wall components. Selectively removing hemicellulose and lignin while reducing crystallinity and increasing biomass surface area can expose cellulose to reducing enzymes, which in turn degrade these larger carbohydrates to readily fermentable mono- and disaccharides (Brodeur et al., 2011). Without pretreatment, the recalcitrant nature of lignocellulose severely limits the ability of hydrolytic enzymes to access the complex carbohydrates thereby reducing production efficiency. Several factors are reported to contribute to the recalcitrant nature of lignocellulosic biomass. These include substrate accessibility, the nature (i.e., amorphous or crystalline), and degree of cellulose polymerization; particle size; porosity; and the percent composition and types of interactions between cellulose, hemicellulose, and lignin (Zhang and Lynd, 2004; Chandra et al., 2007; Himmel et al., 2007). In particular, substrate accessibility is the feature with the most influence in the overall enzymatic efficiency (Zhang and Lynd, 2006; Hong et al., 2007; Arantes and Saddler, 2010; Rollin et al., 2011; Wang et al., 2012b; Luterbacher et al., 2013). Pretreatment alters the cell wall structure by targeting one or more of these factors and serves as an essential step in rendering the cellulose component of the matrix more susceptible to enzymatic action (Figure 2.3).

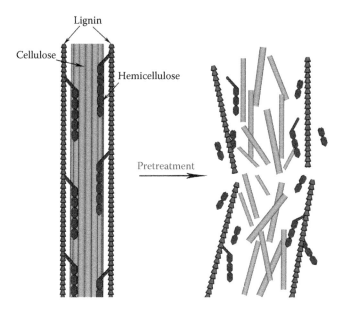

Figure 2.3 Schematic representation of the matrix of polymers in which cellulose exists. Pretreatment of biomass by different methods removes hemicellulose and lignin from this matrix before hydrolysis. (Adopted from Brodeur, G. et al., *Enzyme Res.,* 2011, 787532, 2011.)

Along with enzymatic efficiency, biomass pretreatment efficacy is considered a principal driver in the per-gallon cost of cellulosic ethanol. Pretreatment efficacy impacts all subsequent processing steps, including enzyme-mediated hydrolysis and fermentation operations (Mosier et al., 2005b). Indeed, approximately 30% of the total cost of converting lignocellulosic biomass to ethanol may be associated with pretreatment operations (Lynd et al., 1996). Therefore, improving pretreatment efficacy is a major focus of biofuel research. The most effective pretreatments ultimately reduce the amount of enzyme required in subsequent hydrolysis steps and reduce the energy costs associated with neutralizing batch (or inline) slurries to temperatures or pH values that optimize the catalytic efficiency of hydrolyzing enzymes.

Pretreatments may be considered as either traditional pretreatments or advanced pretreatments. *Traditional pretreatments* for cellulosic biomass include physical or chemical pretreatments, combinational pretreatments, and biological pretreatments (Maurya et al., 2015). With traditional pretreatments, the general goal remains the same, namely disrupting lignin and hemicellulose to improve enzyme access to cellulose for efficient hydrolysis (Pingali et al., 2010). In addition, some pretreatments can decrease the crystalline structure and degree of polymerization of the cellulose in the matrix. Currently, pretreatment options that employ (often hazardous) chemicals are the most efficient (Bensah and Mensah, 2013). However, several novel advanced pretreatments, generally called lignocellulose fractionation pretreatments, may provide new opportunities for efficiently deconstructing cellulosic feedstock. The advanced pretreatment approach can be divided into acid-based fractionation and ionic liquid-based fractionation (ILF)—options. Fractionation methods employ cellulose-specific solvents that not only *loosen up* or dissolve the matrix but also separate macromolecular components into recoverable fractions (Sathitsuksanoh et al., 2013). Although advanced pretreatment technology is considered as one of the top two targets for reducing bioethanol production costs (Mosier et al., 2003a, 2003b), at present, traditional pretreatments are almost exclusively employed for commercial cellulosic ethanol production.

2.3.1   Traditional pretreatments for lignocellulosic biomass

Traditional pretreatment as a first step in deconstructing lignocellulose generally employs physical, chemical, combination (physicochemical), or biological processes.

2.3.1.1   Physical pretreatments

Physical pretreatments do not include chemical agents. The main goal with physical pretreatment is to increase the surface area as well as the size of pores of the biomass. Often physical pretreatments will also decrease cellulose crystallinity and level of polymerization. Physical pretreatment methods include mechanical comminution (reduction of particle size), irradiation, and extrusion (whereby the materials are subjected to heating, mixing, and shearing). Alone, physical pretreatment is not comparatively effective to other methods, and costs are high (Millett et al., 1976, 1979; Horton et al., 1980; Tassinari et al., 1980, 1982; Gracheck et al., 1981; Khan et al., 1987; Rivers and Emert, 1987). Utilization of physical pretreatments is typically conducted in combination with chemical pretreatments to improve deconstruction efficiency.

Mechanical comminution of lignocellulosic materials via chipping, grinding, shredding, milling, or similar processes has been used to enhance the digestibility of lignocellulosic biomass. Mechanical comminution can be effective in decreasing cellulose crystallinity and the degree of polymerization, thereby increasing the available surface area of cellulosic biomass so that the substrate is more susceptible to subsequent enzymatic hydrolysis. Despite these advantages, mechanical comminution is time consuming, energy intensive, expensive, and less effective than chemical pretreatments. Furthermore, it does not remove lignin, which restricts enzymes from accessing cellulose.

Irradiation with high-energy radiation such as γ rays (Yang et al., 2008; Yoon et al., 2012), ultrasonic radiation (Yang et al., 2012; Subhedar et al.,

2015), focused electron beams (Shin and Sung, 2010; Bak, 2014), pulsed electrical fields (Kumar et al., 2011b), UV (Dunlap and Chiang, 1980), and microwaves (Choudhary et al., 2012; Ninomiya et al., 2014) have all been used as another form of physical pretreatment for degrading lignocellulosic biomass. Irradiation can increase the specific surface area, decrease cellulose crystallinity and polymerization, hydrolyze hemicellulose, and polymerize lignin. However, irradiation is typically a slow process that is both energy intensive and expensive (Chang et al., 1981). Furthermore, irradiation methods are substrate specific (Dunlap and Chiang, 1980). Ultimately, high-energy radiation methods are not cost-effective.

Extrusion, however, is a more recent form of physical pretreatment whereby the biomass is subjected to continuous heating, mixing, and shearing. Upon passing through an extruder, the resulting physical and chemical modifications to the lignocellulosic biomass increase accessibility of carbohydrates to enzymatic action (Zhan et al., 2006). Several features of extrusion methods make it an effective and promising form of pretreatment. For example, process temperatures during extrusion are lower than those used in steam explosion (193°C–230°C) reducing premature degradation of complex carbohydrates and oxidation of lignin, which can generate inhibitors to fermentation, or the possibility of continuous operation (de Vrije et al., 2002).

2.3.1.2   Chemical pretreatments

Chemical pretreatments are generally designed to disrupt and remove lignin and/or hemicellulose from the matrix. Some chemical treatments can disrupt the crystalline structure of cellulose; however, this is not generally the case. Chemical pretreatments include acid, alkali, oxidative delignification, and organic acid (organosolvation)—methods. In general, chemical pretreatments are highly selective for specific feedstock types. In most cases, chemical pretreatments are efficient but they often involve harsh reaction conditions, which may have negative effects on downstream processing and require special disposal procedures for processing by-products.

Acid wash is the oldest and most common pretreatment. It involves washing or soaking feedstock with concentrated or dilute acid (or sequences of acid washes) at temperatures from 130°C to 210°C. This disrupts the structural integrity of most lignocellulosic materials. One of the principal actions of acid pretreatment is to degrade hemicellulose especially, xylans. The breakdown of hemicellulose results in two favorable outcomes. First, breaking down hemicellulose releases cellulose, making it more accessible to cellulases and other cellulose degrading enzymes (Pedersen et al., 2011). Second, hemicellulose itself is composed of glucose and other fermentable sugars that may be released via acid wash. The most commonly used acid for such pretreatments is dilute sulfuric acid (H_2SO_4), which has been used in commercial operations for the pretreatment of a broad type of lignocellulosic material (Wyman et al., 2009; Digman et al., 2010; Du et al., 2010; Shuai et al., 2010; Jung et al., 2013). However, several other acids have also been tried, including hydrochloric acid (HCl) (Wang et al., 2010), nitric acid (HNO$_3$) (Kim et al., 2015a), and phosphoric acid (H$_3$PO$_4$) (Isholaa et al., 2014). Concentrated acid is not typically used in ethanol production because of its toxicity, corrosive properties, the high costs associated with recovery, degradation of monosaccharides, and the production of fermentation inhibitors (Sivers and Zacchi, 1995; Pedersen et al., 2010). Thus, the high operational and maintenance costs negate any advantages gained by using concentrated acid pretreatments on a commercial scale (Wyman, 1996). In the end, the use of concentrated acid pretreatment is simply not feasible economically (Sivers and Zacchi, 1995). Dilute acid pretreatments, however, can produce desirable results without the severity of concern regarding acid storage, recovery, and disposal. Dilute acid pretreatments can successfully be employed to break down hemicellulose either as a stand-alone process or in conjunction with other operations as part of a fractionation procedure (Diedericks et al., 2012). Often, an acid pretreatment, which degrades hemicellulose, is followed by alkaline pretreatment, which removes lignin to produce a fairly pure cellulose product (Kim et al., 2012).

Alkaline pretreatment, which is also commonly used, involves soaking the lignocellulosic biomass in alkaline solutions such as sodium, potassium, calcium, or ammonium hydroxides, and mixing it at normal temperature and pressures for a defined period of time. The main effect of this pretreatment is lignin degradation. However, cellulose *swelling* can also occur resulting in: an increase in internal surface area of the matrix, a decrease in the degree of polymerization and crystallinity, and the separation of linkages between lignin and carbohydrates (Fan et al., 1982).

Alkaline pretreatment can also induce partial dissolution of hemicellulose because the process removes acetyl and uronic acid groups present in hemicellulose (Chang and Holtzapple, 2000; Monlau et al., 2013). The most commonly used alkaline pretreatments employ calcium hydroxide (lime) washes or sodium hydroxide washes. Reaction conditions during alkaline pretreatments are usually milder (lower temperatures, pressures, and residence times) than other pretreatment methods, and the degree is directly dependent on the nature of the biomass feedstock, in particular its lignin content (McMillan, 1994; Mirahmadi et al., 2010). Mild reaction conditions reduce unwanted breakdown of desired sugars (Sharma et al., 2013b). The mild reaction conditions also prevent condensation of lignin, which leads to increased lignin solubility (and more effective lignin removal). Alkaline pretreatments can be combined with air or oxygen input, which improves the lignin degradations process (Chang and Holtzapple, 2000). Lime pretreatment has been shown to disrupt amorphous elements of the matrix, such as lignin. This prepares the material for enzymatic breakdown by reducing nonproductive adsorption sites to which enzymes may bind and increase cellulose accessibility (Kim and Lee, 2006). Lime pretreatment also enhances the efficiency of subsequent enzymatic processes by probably reducing steric interference through the removal of acetyl groups, thereby improving hemicellulose and cellulose digestibility (Chang and Holtzapple, 2000). Sodium hydroxide (NaOH) pretreatment swells or *loosens* the matrix, thereby increasing the internal surface area of the cellulose structure, decreasing polymerization, reducing cellulose crystallinity, and disrupting lignin interactions and structure (Fan et al., 1982, 1987). Ammonia pretreatment is a special form of alkaline pretreatment that uses aqueous ammonia at elevated temperatures. As seen with other pretreatments, ammonia pretreatment targets lignin; however, hemicellulose and crystallinity within the cellulose structure are also disrupted. Lignocellulosic materials with lower lignin content respond favorably to ammonia pretreatment. The two major drawbacks of using ammonia are the cost and the environmental concerns. Ammonia pretreatments include three types of procedures: soaking in aqueous ammonia (SAA) (Kang et al., 2012), ammonia recycle percolation (ARP) (Kim et al., 2003), and the ammonia fiber explosion method, commonly referred to as, AFEX (Teymouri et al., 2005). SAA pretreatment is performed at low temperatures and efficiently removes lignin from the matrix by disrupting interactions between lignin and hemicellulose (Qin et al., 2013). Using SAA, hemicellulose is retained as a solid and is easily removed. This is advantageous because it averts a separate recovery step to remove xylose from the slurry (Kim and Lee, 2005a). ARP pretreatment employs a flow-through column reactor, and the ammonia is recovered from the wash. Specifically, the aqueous ammonia (5–15 wt %) at elevated temperatures (150°C–180°C) is passed through a reactor column packed with biomass at 1–5 mL/min for 10–90 min (Kim et al., 2003; Kim and Lee, 2005b, 2006; Ramirez et al., 2013). ARP hydrolyzes hemicellulose and destroys lignin by ammonolysis. ARP pretreatment also breaks hydrogen bonds that disrupt the crystalline structure of cellulose, making cellulose more susceptible to enzymatic action (Schuerch, 1963; Mittal et al., 2011).

The conventional AFEX pretreatment consists of treating lignocellulosic biomass with liquid anhydrous ammonia (0.3–2 g NH_3/g dry biomass) at elevated temperature (40°C–180°C) and pressure (250–300 psi) for 5–60 min, then rapidly reducing the pressure to facilitate expansion of the ammonia gas (Balan et al., 2009). This induces swelling in lignocellulosic matrix, disruption in the lignin–carbohydrate linkage, hemicellulose and lignin hydrolysis, ammonolysis of glucuronic cross-linked bonds, and cellulose decrystallization (Laureano-Perez et al., 2005; Chundawat et al., 2007). Although lignin is not robustly affected during the process, it has been reported that close to 100% of the cellulose obtained after AFEX pretreatment can be converted to fermentable sugars via enzymatic action (Teymouri et al., 2005; Balan et al., 2008). Moreover, ~100% of the ammonia can be recovered or removed, and AFEX does not result in the formation of downstream inhibitors to subsequent biological processes (e.g., fermentation) (Dale and Moreira, 1983; Srebotnik et al., 1988). However, the more outstanding results from AFEX may be limited to feedstock that is derived from agricultural residues and herbaceous crops. Efficacy is limited on materials with high lignin (McMillan, 1994). Both AFEX and ARP have only been reported in lab-scale use. AFEX used in conjunction with other methods may also be considered an *advanced* pretreatment technology (Section 2.3.2).

Oxidative delignification pretreatment involves the addition of an oxidizing compound such as hydrogen peroxide, ozone (ozonolysis), oxygen, or air (wet oxidation) to the biomass slurry (Hammel et al., 2002; Arvaniti et al., 2012; Li et al., 2015). Despite the favorable matrix deconstruction outcomes, oxidative delignification is a costly process and generates simpler acids, which have to be neutralized or removed because they inhibit fermentation. Hemicellulose may also be degraded becoming unavailable for fermentation (Ghedalia and Miron, 1981; Li et al., 2015). Hydrogen peroxide (H_2O_2) is the most common oxidizing agent used in oxidative delignification. H_2O_2 reacts with iron via the Fenton reaction to generate hydroxyl radicals, which degrades lignin, producing low molecular weight products (Hammel et al., 2002). As mentioned earlier, removal of lignin from lignocellulose leads to greater exposure of cellulose and hemicellulose to enzymes, resulting in improved hydrolysis (Ding et al., 2012). Reducing sugar yields can be as high as 90% when H_2O_2 is combined with NaOH pretreatment (Cao et al., 2012). Alternatively, ozone (O_3) may be used to induce ozonolysis. Ozone is a powerful oxidant that breaks down lignin by attacking aromatic rings structures or the side chains of the hydroxycinnamyl alcohols. Ozonolysis reduces lignin content in lignocellulosic biomass and hemicellulose while leaving cellulose nearly intact (Linder et al., 2005; Li et al., 2015). This has been shown to be an effective method for deconstructing biomass feedstock such as wheat straw, sugarcane bagasse, maize stover, and energy grasses (Binder et al., 1980; García-Cubero et al., 2009; de Barros et al., 2013; Travaini et al., 2013; Panneerselvam et al., 2013a, 2013b; Li et al., 2015). Ozonolysis is most robust when used with other pretreatment processes (de Barros et al., 2013). This approach does not generate inhibitory compounds; however it requires large amount of ozone, which makes the process economically unviable for large-scale commercial applications (Alvira et al., 2010).

Wet oxidation uses oxygen or air in combination with water at elevated temperature and pressure (McGinnis et al., 1983; Arvaniti et al., 2012). The mechanism implies autocatalyzing by formed organic acids from hydrolytic processes and oxidative reactions (McGinnis et al., 1983). Wet oxidation decrystallizes cellulose and solubilizes hemicellulose and lignin (McGinnis et al., 1983; Martín et al., 2007; Banerjee et al., 2009). Hemicellulose is solubilized, and lignin is degraded into carbon dioxide, water, and carboxylic acids such as glycolic acid, formic acid, isobutyric acid, and acetic acid (Bjerre et al., 1996). Wet oxidation pretreatment can result in reducing sugar yields up to 70% (Banerjee et al., 2009). The primary disadvantage of wet oxidation pretreatment is that it requires high temperature, high pressure, oxygen, and catalysts that limit its economic viability.

Organosolvation pretreatment uses a variety of organic or aqueous solvent mixtures, which may include methanol, ethanol, acetone, and/or ethylene glycol to solubilize lignin and hemicellulose (Zhao et al., 2009). Organosolvation pretreatment (a.k.a., Organosolv) may be performed with or without a catalyst (Quesada-Medina et al., 2010; Amiria et al., 2014). When used with a catalyst, the degradation of hemicellulose and the overall digestibility of the cellulosic biomass are enhanced (Chum et al., 1988). Normal operating temperatures for organosolv are between 150°C and 200°C (Rajendran and Taherzadeh, 2014) (Figure 2.4). For more

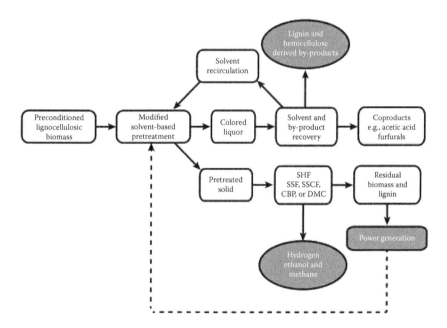

Figure 2.4 A schematic outline of a multiproduct biorefinery platform based on a modified solvent-based pretreatment capable of selectively purifying cellulose, solubilizing hemicelluloses, and precipitating lignin; for example, organosolv pretreatment. SHF, separate hydrolysis and fermentation; SSF, simultaneous saccharification and fermentation; SSCF, simultaneous saccharification and cofermentation; CBP, consolidated bioprocessing; DMC, direct microbial conversion. Dotted arrow represents potential power generation options available to plants producing excess amounts of biofuels and residual biomass that can be used to generate power to run the plant. (Adopted from Agbor, V.B. et al., *Biotechnol. Adv.*, 29, 675–685, 2011.)

recalcitrant materials, organosolv can be used in conjunction with other pretreatment methods to obtain a more digestible end product ready for enzyme-mediated sugar reduction (Rughani and McGinnis, 1989; Itoh et al., 2003; Hongzhang and Liying, 2007). Organosolvation techniques have several advantages. First, organosolv generates a fairly pure lignin by-product (Cybulska et al., 2012). Second, organosolvation requires less energy compared to other pretreatment methods. For example, it does not require extensive biomass particle size reduction (Pan et al., 2005; Silverstein et al., 2007; Zhu et al., 2009a). Despite the advantages, organosolv is costly and generates by-products that can significantly inhibit enzyme action and fermentation. The solvents used during the process must be removed and recovered via evaporation and condensation procedures (Zhao et al., 2009). Still, this technology has been used for various lignocellulosic feedstocks including softwoods, hardwoods, agroenergy crops, and agricultural residues (Pan et al., 2006, 2008; Brosse et al., 2009; Hallac et al., 2010; Sannigrahi et al., 2010a).

2.3.1.3 *Physicochemical pretreatments*

Physicochemical pretreatments feature conditions and compounds that target both physical and chemical properties of biomass. They include steam explosion, liquid hot water, microwave irradiation, and CO_2 explosion pretreatments. ARP and AFEX can be incorporated into these pretreatments (Alvira et al., 2010; Brodeur et al., 2011).

Steam pretreatment is the most widely studied and the most frequently employed physicochemical method for any lignocellulosic material (Chandra et al., 2007; Balat, 2011; Wanderleya et al., 2013; Singha et al., 2015). It involves heating the biomass at high temperatures and pressures followed by either a quick decrease in pressure (explosion) or a slow decrease in pressure (no explosion) (Brownell and Saddler, 1987; Shamsudin et al., 2012; Shafiei et al., 2015). In particular, steam explosion pretreatment employs high pressure (0.7–4.8 MPa) saturated steam for 30 s to 20 min at temperatures of 160°C–260°C followed by a sudden reduction in pressure (Agbor et al., 2011). It was originally thought that steam explosion pretreatment worked via mechanical force and chemical effects; however, it is now known that the chemical effects are the principal mode of action.

Although separation of fibers may occur in limited fashion as a result of the explosive decompression, it is the chemical modifications within lignocellulose that have the greatest impact on the material (Brownell and Saddler, 1987). Specifically, steam explosion pretreatment solubilizes hemicellulose by autohydrolysis (if no exogeneous acid catalyst is added) via the conversion of acetyl groups to acetic acid. Other acids may also be generated, ultimately leading to hemicellulose hydrolysis (Ramos, 2003).

The major change during steam explosion pretreatment is the hydrolysis of hemicelluloses. However, other changes such as modification and degradation of the lignin, increasing the crystallinity, and decreasing the polymerization of cellulose are also reported (Jeoh, 1998; Ramos, 2003; Garmakhany et al., 2013). Steam explosion pretreatment can be enhanced by adding H_2SO_4, CO_2, or SO_2 as a catalyst. The addition of catalyst decreases the temperature and time to improve enzymatic hydrolysis, reduces the production of compounds that inhibit downstream processes, and leads to more complete removal of hemicellulose (Stenberg et al., 1998). Steam explosion pretreatment, when acid catalyst is not used, is an environment-friendly method because it does not use chemical reagents (Egüés et al., 2012). Unlike some pretreatments, there is a limited use of industrial chemicals, low energy requirements, and low capital investment while achieving a robust sugar yield (Avellar and Glasser, 1998; Sawada and Nakamura, 2001; Conde-Mejía et al., 2012; Egüés et al., 2012). Disadvantages of steam explosion pretreatments include partial xylan and lignin degradation, incomplete disruption of the lignin–carbohydrate matrix, lignin redistribution on the cellulose surface, and the generation of toxic compounds that affect enzymatic hydrolysis and fermentation (Mackie et al., 1985; Ramos et al., 1992; Ramos and Saddler, 1994; Emmel et al., 2003; Zhang et al., 2013).

Liquid hot water pretreatment is similar to steam explosion pretreatment. However, instead of using steam and rapid decompression, constant high pressure (>5 MPa) is used to maintain water in the liquid state at elevated temperatures (170°C–230°C) (Sanchez and Cardona, 2008). Liquid hot water pretreatment solubilizes most of hemicellulose, and lignin, enhancing cellulose accessibility while avoiding the generation of inhibitors to fermentation (Kohlmann et al., 1995; Weil et al., 1998; Yang and Wyman, 2004; Mosier et al., 2005a). With liquid hot water pretreatment, there is no need for costly reactors. Acids are liberated from biomass by the cut of hemiacetal linkages by the hot water, which break the ether linkages in biomass (Antal Jr., 1996). Generally, pretreatment with hot water is inexpensive compared to other pretreatment options. The low-corrosion potential and the fact that neither final washing nor slurry neutralization steps are required make this type of pretreatment attractive. Furthermore, there are few to nil by-products that serve as inhibitors to subsequent biological processes and no chemicals are generally required (Mosier et al., 2005a; Kim et al., 2009a; Xu et al., 2010; Li et al., 2014). Despite such advantages, liquid hot water pretreatment does require large amounts of water. There is also a significant energy demand associated with maintaining high-pressure environments for long periods of time. This pretreatment has shown promise on a lab scale, and pH-controlled hot water pretreatment has been applied to large-scale pretreatment of corn fiber (Mosier et al., 2005c).

Microwave irradiation has also been employed as a pretreatment of lignocellulosic biomass (Ooshima et al., 1984; Choudhary et al., 2012). Microwaves are electromagnetic radiation with wavelengths ranging from 1 m to 1 mm and with frequencies ranging between 300 MHz and 300 GHz. As previously mentioned, microwaves can have a physical impact on cellulosic biomass. At selective wavelengths and pulse frequencies, matrix chemistry can be more readily altered resulting in physical and chemical modifications to lignocellulose. Physical effects arise in part from the generation of internal heat within the biomass via polar bond vibrations resulting from microwave irradiation (Watkins, 1983; Neas and Collins, 1988; Mingos and Baghurst, 1991). This internal heating effect disrupts the physical structure of lignocellulose, loosening the matrix and enhancing exposure of cellulose and hemicellulose to enzymatic hydrolysis (Hu and Wen, 2008). Chemical changes occur as a result of thermally treating lignocellulose in an aqueous medium. This results in the release of acetic acid and provides an acidic environment where autohydrolysis occurs (Lora and Wayman, 1978). Microwave irradiation pretreatment results in alteration of the ultrastructure of cellulose, removal of lignin and hemicellulose in lignocellulose, and enhancement of the enzymatic susceptibility of lignocellulosic materials (Azuma et al., 1984; Xiong et al., 2000; Hu and Wen, 2008). Reducing sugar yields after microwave irradiation is considered reasonable for industrial application with ranges from ~40% to ~60% (Verma et al., 2011b). Still, the capital costs required to purchase and operate a large microwave reactors make this pretreatment costly for large-scale operations (Martin, 2009). Furthermore, the high temperatures involved often result in a heating effect on the biomass that is not uniform, which can confound efficiency expectations. Microwave pretreatment can also result in the formation of inhibitors to fermentation processes (Jackowiak et al., 2011). These disadvantages limit the use of microwave irradiation in large-scale bioethanol production facilities.

CO_2 explosion pretreatment employs carbon dioxide (CO_2) as a supercritical fluid (supercritical fluid refers to a fluid that exists under select conditions as either a liquid or gas). Above the critical temperature and pressure (e.g., at the critical point) in which gas and liquid coexist, supercritical CO_2 permeates the biomass in a high-pressure vessel (Kim and Hong, 2001). In some processes, high pressures (1000–4000 psi) and elevated temperatures (up to 200°C) can be alternately applied for a defined time period to achieve maximum CO_2 penetration (Puri and Mamers, 1983; Zheng et al., 1995). Upon dissolving the treated cellulosic biomass in water, carbonic acid is produced, which specially facilitates the hydrolysis of hemicellulose (Puri and Mamers, 1983; Puri, 1984; Meyssami et al., 1992; Kim and Hong, 2001; van Walsum and Shi, 2004). Upon reduction of pressurization with CO_2, the structure of the biomass is altered such that there is a significant increase in the accessible surface area and in some cases a reduction in the degree of cellulosic crystallinity, a desirable outcome for subsequent enzymatic hydrolysis (Zheng et al., 1995, 1998; Narayanaswamy et al., 2011).

Compared to other chemical pretreatments, CO_2 explosion pretreatment is a low cost alternative method because the carbon dioxide solvent is readily available and inexpensive. Moreover, CO_2 explosion does not generate strong inhibitors to subsequent biological processes (Srinivasan and Ju, 2010; Islam et al., 2014). The use of low temperatures and high solid capacity makes this approach reasonable from an energy expenditure perspective. However, the cost of the equipment required for commercial scale production has prevented the wide-scale adoption of this technique.

2.3.1.4   Biological pretreatments

Beyond standard (and experimental) physical, chemical, and physico-chemical pretreatments, several biological pretreatment methods have been tested. Most of these biological pretreatments employ wood degrading microorganisms to modify the chemical composition and/or structure of the lignocellulosic feedstock. This can be a precursor to using one or more of the other aforementioned pretreatments or, in some cases, can be used as a stand-alone pretreatment option. Microorganisms used for biological pretreatment include white-rot fungi, brown-rot fungi, soft-rot fungi, and bacteria that express lignocellulose-degrading enzymes (Schurz, 1978; Kurakake et al., 2007). The general approach is to utilize these microbes to initiate the breakdown of the matrix so that subsequent enzymatic hydrolysis of cellulose proceeds with greater efficiency. However, each microbe is selective for different components of the matrix. For example, brown-rot fungi selectively deconstruct cellulose and hemicellulose with virtually no modification of lignin. White-rot fungi act actively on lignin, along with cellulose and hemicellulose. Soft fungi act on polysaccharides and may also degrade lignin (Hatakka and Hammel, 2010). Interestingly, it appears that white-rot fungi are the most effective microorganisms in biological pretreatment (Fan et al., 1987; Yu et al., 2009). However, new biological pretreatments, which employ novel or genetically modified microbes may prove effective as such pretreatment technologies are further developed. In general, biological pretreatments feature low capital costs, low energy input, no hazardous chemical requirements, and mild environmental conditions. However, the slow rates of hydrolysis and thus requirements for long pretreatment times are major drawbacks for this technology (Maurya et al., 2015). Furthermore, each biological pretreatment type is feedstock specific; there is no general biological pretreatment that works efficiently on all types of lignocellulosic material. Current research and development in biological pretreatment is focused on combining microbial degradation as a *pre*-pretreatment to other commonly used pretreatments (Ma et al., 2010; Wang et al., 2012d).

More advanced studies are focused on discovering novel lignocellulose-degrading microbes or on producing genetically modified microorganisms with an ability to selectively disrupt lignin and to release usable carbohydrates (Canam et al., 2011; Zambare et al., 2011).

2.3.2   Advanced pretreatments for lignocellulosic biomass

Pretreating lignocellulosic biomass to enhance the efficiency of subsequent enzymatic hydrolysis and fermentation processes is the key to ensuring the economic viability of cellulosic ethanol. Clearly, there is not a single technology that maximizes end product yields with every feedstock. Each method currently in use or under development has distinct advantages and disadvantages on selected feedstock types. Therefore, new methods are constantly being developed to maximize cellulose accessibility, to generate convertible products from hemicellulose, and to remove lignin. As described earlier, common pretreatment alternatives include acid wash (Jung et al., 2013), alkaline wash (Sharma et al., 2013b), high temperature steam explosion (Garmakhany et al., 2013), and mixed washes including AFEX with dilute acid. Until recently, AFEX was considered an advanced pretreatment. Indeed, AFEX with dilute acid is one of the leading alkaline pretreatment technologies currently used in industry due to its ability to enhance enzyme-mediated cellulose reduction without the need for procedures that physically remove hemicellulose and lignin from the feedstock. The key to AFEX is that it reduces the competitive binding of cellulases to xylan (and xylooligomers) from hemicellulose, enhancing the enzymatic reduction of cellulose (Qing and Wyman, 2011). (Recall that hemicellulose not only stabilizes the lignocellulosic matrix but that the xylan within hemicellulose can also serve as a competitive binding site for cellulose degradation enzymes.)

More recently, efforts have been undertaken to minimize costs in cellulosic ethanol production by fractionating the lignocellulosic biomass in a manner that produces value-added coproducts (de Jong et al., 2012). A newer class of advanced pretreatments collectively called lignocellulose fractionation pretreatment is based on this idea. The aim of these new pretreatment methods is to employ cellulose solvents not only to improve cellulose accessibility for subsequent enzymatic hydrolysis but also to separate the principal constituents of cellulosic materials (i.e., cellulose, hemicellulose, and lignin) to produce value-added coproducts (Sathitsuksanoh et al., 2013). Specialized cellulose solvents are designed to increase cellulose accessibility by dissolving lignocellulose biomass so that subsequent enzymatic treatments are effective at low enzyme loading levels. Furthermore, solvents are developed to facilitate fractionation of the primary components (i.e., cellulose, hemicellulose, and lignin) under modest reaction conditions such as atmospheric pressure and 50°C (Ladisch et al.,

1978; Zhang et al., 2007). The most effective solvents deconstruct lignocellulose, selectively separate primary components, and generate high yields of quality coproducts while minimizing overall costs of processing (Van Heiningen, 2006; Wising and Stuart, 2006). Cellulose solvent-based lignocellulose fractionation (CSLF) is a set of cutting-edge technologies that are more efficient than traditional biomass pretreatments such as steam explosion (Garmakhany et al., 2013), AFEX (Shao et al., 2011), SAA (Kang et al., 2012), dilute acid pretreatment (Zhu et al., 2009b), or organosolv (Amiria et al., 2014).

For most CSLF methods, hydrolysis rates and digestibility of pretreated biomass are improved. Furthermore, the amount of enzyme required in subsequent enzymatic hydrolysis steps is decreased, and multiple feedstock types may be treated (Sathitsuksanoh et al., 2009, 2010, 2012).

Two general approaches to CSLF have been developed: (a) acid-mediated fractionation and (b) ILF. Each has selective advantages for treating common lignocellulosic feedstock.

2.3.2.1 Acid-mediated fractionation

Cellulose solvent and organic solvent lignocellulose fractionation (COSLF) technology has separated lignocellulose components using cellulose solvent such as phosphoric acid and an organic solvent (e.g., acetone or ethanol) under modest reaction conditions (1 atm, 50°C) (Ladisch et al., 1978; Zhang et al., 2007; Rollin et al., 2011). The successful fractionation of lignocellulosic biomass depends upon the solubility characteristics of cellulose, hemicellulose, and lignin in the cellulose solvent, organic solvent, and water, respectively (Zhang et al., 2007). Concentrated phosphoric acid disrupts the linkage among cellulose, hemicellulose, and lignin (Moxley et al., 2008); dissolves cellulose fibers disrupting the highly ordered hydrogen bonds of the cellulose crystalline structure (Zhang et al., 2006a; Sathitsuksanoh et al., 2011); and significantly increases enzyme access to cellulose fibers (Zhu et al., 2009b; Rollin et al., 2011). Generally, the cellulose solvent can be recovered/recycled as a part of the process due to the difference in volatility with the organic solvents (Zhang et al., 2007). Strategic sequencing and timing of cellulose solvent, organic solvent, and water treatments have resulted in significant decreases of crystallinity (Zhang et al., 2006a, 2007); removal of lignin and hemicellulose from the cellulose fraction, which reduces substrate obstacles and competitive binding sites; reduction of unwanted sugar degradation; reduction of the inhibitors production to subsequent biological processes; and reduction of utility consumption and capital investment (Zhang et al., 2007). COSLF has been shown to efficiently pretreat different types of lignocellulosic biomass such as bamboo (Sathitsuksanoh et al., 2010), common reed (Li et al., 2009; Sathitsuksanoh et al., 2009), hemp hurd (Moxley et al., 2008), corn stover (Zhu et al., 2009b), bermudagrass (Li et al., 2009), switchgrass

(Sathitsuksanoh et al., 2011), gamagrass (Ge et al., 2012), giant reed, ele-
phant grass, and sugarcane (Ge et al., 2011). Although design of novel cellulose solvent solutions and optimization of wash times and sequencing
can be a challenge, no major disadvantages have been identified in using
COSLF as a commercial-scale lignocellulosic biomass pretreatment.

2.3.2.2   Ionic liquid-based fractionation

ILF is another type of advanced pretreatment based on CSLF. Ionic liquids (ILs) have also been shown to dissolve carbohydrates and lignins
(Zakrzewska et al., 2010; Hossain and Aldous, 2012). ILs can be used for
fractionation of lignocellulosic biomass to obtain specific, purified, and
unaltered polymeric raw materials that can readily be separated and used
as valuable coproducts (Lee et al., 2009). Typically, ILs are salt solutions
composed of large organic cations and small or inorganic anions, which
exist as liquid at relatively low temperatures such as room temperature
(Wasserscheid and Keim, 2000; Rogers and Seddon, 2003). ILs are considered as *green* solvents because of environmentally friendly properties such
as low vapor pressure and high thermal stability (Paulechka et al., 2003;
Kabo et al., 2004; Domanska and Bogel-Lukasik, 2005). This differs from
the properties of classical volatile solvents. ILs are also *tunable*. Properties
such as hydrophobicity, polarity, and solvent power can all be adjusted to
meet specific process needs (Huddleston et al., 2001; Vasiltsova et al., 2004;
Chiappe and Pieraccini, 2005). Similar to COSLF, IL pretreatment disrupts
interactions between lignin–carbohydrate complex, thereby enhancing
the general accessibility of the material for enzymatic hydrolysis (Singh
et al., 2009). After dissolution, a regeneration and fractionation step is performed. Regeneration of solute is done by precipitating in the presence
of an antisolvent such as water, acetone, dichloromethane, or acetonitrile
(Fort et al., 2007). Regeneration is a result of preferential solute displace-
ment. With some antisolvents, up to 80% of lignin and hemicellulose can
be fractionated (Sun et al., 2009; Li et al., 2010, 2011). The regenerated fraction is essentially composed of carbohydrates; however, lignin can be partially extracted in the ionic solvent/antisolvent mixture (Lee et al., 2009;
Sun et al., 2009). Antisolvent is separated from the ionic liquid, and they
both can be reused (Dadi et al., 2006; Zhu et al., 2006). ILs inhibit enzymes
(Turner et al., 2003; Docherty and Kulpa, 2005; Murugesan et al., 2006;
Engel et al., 2010; Salvador et al., 2010). Complete removal of the IL is not
economically viable; therefore, design of IL-tolerant enzymes is a current
research focus (Datta et al., 2010; Bose et al., 2012). In addition, the use
of high pressure during enzymatic reactions could improve enzymatic
activities in the presence of ionic liquids (Salvador et al., 2010). Overall
efficiency of IL pretreatment is determined by the properties of the IL,
lignocellulosic biomass properties (type, moisture content, partial size,
and load), temperature, time of pretreatment, and the properties of the

antisolvent used (da Costa Lopes et al., 2013). IL pretreatment offers several advantages over traditional lignocellulose pretreatments, including significant alteration in the physicochemical properties of the biomass, such as a major reduction of lignin and cellulose crystallinity (Lee et al., 2009; Doherty et al., 2010); extraction of specific matrix components, such as isolation of lignin (Tan et al., 2009) and cellulose (Abe et al., 2010); flexibility in the application of different fractionation strategies after the dissolution of biomass in the IL (Dibble et al., 2011; Lan et al., 2011; Yang et al., 2013); low energy demands; and ease of operation (Zhang et al., 2007; Yoon et al., 2011). In addition, it was widely believed that IL was an environmentally friendly process; however, recent evidence suggests that this may not be the case (Plechkova and Seddon, 2006; Bubalo et al., 2014).

Although technically viable, the IL pretreatment approach is still expensive. Further research is required to optimize IL pretreatments so that they are economically viable and applicable at larger scales.

2.4   Summary: Feedstocks and processing

Simple sugars such as glucose and sucrose are the primary substrates for microbial-based (i.e., yeast) fermentation processes. By fermentation, these simple carbohydrates are converted to ethanol. Fermentation methods have been employed for millennia to produce beer, wine, and other alcoholic beverages. In modern times, ethanol production has been considered a viable alternative liquid fuel that may reduce global dependency on fossil-based fuels such as gasoline. Although there is a natural feedstock from which primary substrate may be extracted, most of these sources (e.g., corn and sugarcane) are used for food creating a *food-versus-fuel* dilemma. Feedstocks, such as corn and sugarcane, from which primary substrates are readily extracted for bioethanol are considered as *first-generation* feedstocks. Efforts to identify alternative feedstocks that do not compete with food supplies and that minimize land and water use have led to the development of *second-generation* feedstocks such as corn stover, bagasse, woody biomass, grasses, crop straws, and other lignocellulosic materials that do not compete with food crops. Ethanol derived from these feedstocks is called cellulosic ethanol. As lignocellulosic biomass is the most abundant natural material on the planet, significant effort has been placed on developing methods to produce cellulosic ethanol in a cost-competitive manner as a supplement or substitute for liquid fossil fuels. However, second-generation feedstock sources do not readily provide simple fermentable sugars. Secondary substrates, mainly cellulose and hemicellulose, are the carbohydrate source from which simpler, fermentable sugars are produced. Conversion of long-chain carbohydrates requires hydrolysis, generally enzyme-mediated hydrolysis, to break down these secondary substrates into primary substrates (e.g., glucose) for fermentation. Unfortunately, cellulose and hemicellulose are tightly bound with lignin (and often pectin) in a recalcitrant lignocellulosic matrix, which is found virtually in all plant cell walls.

In order to provide sugar reducing enzymes, such as cellulases and xylanases, access to the lignin-bound cellulose and hemicellulose, pre-treatments are typically required to loosen up the matrix. This includes breaking interactions between cellulose, hemicellulose, and lignin; increasing the exposed surface area of the cellulose network within the biomass; and optimizing conditions for efficient enzyme action. Traditional pretreatments strategies include washing feedstock in acidic solutions and/or alkaline solutions; physically swelling or expanding the material by steam or high-temperature liquid water perfusions; irradiating feedstock with microwaves, gamma rays, or other electro-magnetic radiation; applying special solvents or ionic solutions; and other methods to separate and expose cellulose (and hemicellulose) to enzyme. Unfortunately, along with the selected advantages, each pre-treatment strategy features disadvantages, which ultimately render the process either inefficient or economically unviable for commercial scale cellulosic ethanol production.

Despite these challenges, new pretreatment techniques collectively referred to as cellulose solvent-based liquid fractionation may provide a cost-effective way to produce bioethanol from lignocellulosic biomass in a manner that is cost competitive with fossil-based liquid fuels. Advances in pretreatment technologies may open the market to feedstock options that have been considered economically unsuitable for cellulosic ethanol production. Efficient pretreatment strategies along with a broader selec-tion of lignocellulosic feedstock will position cellulosic ethanol to a more favorable market position. However, in addition to new, more efficient pretreatment methods and a broader selection of feedstock options, there is another major step within the bioethanol production process that can contribute to producing a cost-competitive ethanol product.

Significant room for improvement exists in the post-pretreatment enzymatic hydrolysis step of the ethanol production process. Optimizing enzymatic action, discovering novel high-efficiency enzymes, designing genetically engineered enzymes, and using engineered platforms are all areas of scientific investigation underway to further lower the per-gallon cost of bioethanol. In Chapter 3, enzymes and enzymatic action on ligno-cellulosic biomass will be addressed with a focus on synergistic action of multienzyme complements and their ability to further break down lignin and convert cellulose (and hemicellulose) to fermentable sugars.

chapter three

Natural enzymes used to convert feedstock to substrate

3.1 Mode of action of primary lignocellulolytic enzymes

To use cellulosic material as feedstock for bioethanol production, chipped or ground biomass is typically pretreated to facilitate enzyme access to long-chain carbohydrates (e.g., cellulose), which are the macromolecules that are reduced to fermentable sugar for conversion to ethanol. Given the heterogeneous nature of lignocellulose, it is highly recalcitrant even with pretreatment. Numerous methods have been developed for degrading lignocellulose to expose polysaccharides. These include soaking feedstock in acidic or alkaline solution and mechanical disruption. Numerous methods have also been developed to reduce these polysaccharides to simple sugars. Methods that rely on cellulolytic enzymes that are derived from microorganisms can be highly efficient. Both multidomain enzymes and enzyme complexes (e.g., minicellulosomes) have been applied. One useful feature of many lignocellulolytic enzymes (and their complexes) is innate modularity. In addition to a catalytic core region, many cellulolytic enzymes possess noncatalytic domains. Two notable domains include carbohydrate-binding modules (CBMs) and dockerin domains. CBMs facilitate interactions between enzymes and their respective carbohydrate substrates (Tomme et al., 1988a, 1998b; Boraston et al., 1999; Gilbert et al., 2013). Various studies have demonstrated that CBMs enhance enzymatic activity against recalcitrant substrates (Black et al., 1996; Bolam et al., 1998; Carrard et al., 2000; Mello and Polikarpov, 2014). Although three-dimensional structures and specific modes of action may vary between CMBs from different species, in general, CMBs facilitate binding between large enzymatic complexes and substrates so that the probability of enzymatic action (i.e., substrate cleavage) is increased. Dockerin domains on cellulolytic enzymes from some species of microorganisms mediate cohesin–dockerin interactions, associating the enzymes with larger macromolecular complexes. These complexes, or *cellulosomes*, are found naturally in the cell membrane and cell wall structure of many cellulolytic microorganisms (Fontes and Gilbert, 2010). Cellulosome structure and function can vary between microorganisms expressing cohesin and dockerin-based complexes; however, a general

feature of most cellulosomes is the ability to alter the enzyme complement within the complex depending on the type of substrate available in
the environment. Microorganisms will have characteristic repertoires of
lignocellulolytic enzymes that may be expressed at different levels and
incorporated into the cellulosome depending on the environmental conditions and the metabolic needs. Lignocellulolytic enzymes are categorized
as cellulases, hemicellulases, ligninolytic enzymes, and pectinases. In this
section, a review of primary cellulolytic enzymes and their respective functions within natural cellulosomes is provided.

3.1.1   Cellulases

Cellulases are glycosyl hydrolases (GHs) that catalyze cellulolysis—the
cleaving of glycosidic bonds in cellulose. The enzyme-mediated cleavage
of β-1,4-glycosidic bonds in cellulose occurs via acid hydrolysis, using a
proton donor and a nucleophile or base. The products of acid hydrolysis
either result in an inversion or retention through single or double replacement, respectively, of the anomeric configuration of the carbon-1 (C-1) at
the substrate-reducing end (Koshland, 1953; Vocadlo and Davies, 2008). In
macromolecular complexes, such as in naturally occurring cellulosomes,
a complement of cellulolytic enzymes typically acts in a synergistic
manner (Wood and McCrae, 1979; Lamed et al., 1983b; Fierobe et al., 2001).
Cellulolytic enzyme synergism can be measured qualitatively and quan-
titatively; however, predicting synergistic effects of novel combinations of
enzymes either free in solution or bound in an artificial cellulosome has
proven challenging and is the subject of intense investigation. According
to the studies on fungi (Selby and Maitland, 1967; Wood and McCrae,
1972, 1978; Berghem et al., 1976; Mandels and Reese, 1999), bioconversion
of polysaccharide substrates into simple fermentable sugars requires synergistic interactions of at least three types of enzymes: endoglucanases,
cellobiohydrolase, and β-glucosidases. Most of these components are glycoproteins, and each presents isoenzymes in natural systems (Wood and
McCrae, 1972; Gilkes et al., 1984; Mihoc and Kluepfel, 1990; Jimenéz-Zurdo
et al., 1996; Igual et al., 2001; Wei et al., 2005; Begum and Absar, 2009;
Khalili et al., 2011). Functionally, cellulases may be generally categorized
into three groups based on the type of reaction that they catalyze: carbohydrases (including, endoglucanases, exoglucanases, and cellobiases),
oxidative cellulases (e.g., cellobiose dehydrogenase [CDH]) and phosphorylases (i.e., cellobiose phosphorylase and cellodextrin phosphorylase).

3.1.1.1   Carbohydrases

Carbohydrases are glycosyl hydrolases that hydrolyze the β-1,4-glycosidic
bonds of cellulose or cellooligosaccharides, leading to the formation of
short cellodextrins and molecules of glucose (Lombard et al., 2014; CAZy,

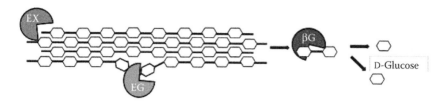

Figure 3.1 Schematic graphic of cellulose structure and mode of action of cellulo-lytic enzymes EX, EG, and βG, leading to the formation of D-glucose. (Adopted from van den Brink, J. and de Vries, R.P., *Appl. Microbiol. Biotechnol.*, 91, 1477–1492, 2011.)

2015). There are three types of carbohydrases: endoglucanases or *endocellulases* (EGs), exoglucanases or *exocellulases* (EXs), and β-glucosidases (βGs) also referred to as β-D-glucoside glucohydrolases or *cellobiases*.

EGs (e.g., EC 3.2.1.4) are 1,4-β-D-glucan-4-glucanhydrolases that disrupt bonds at random internal sites in the cellulose polysaccharide chain, producing oligosaccharides of various lengths. More generally referred to as endocellulases, the EGs produce new chain ends (Figure 3.1). EGs that do not feature CBMs hydrolyze at amorphous internal sites within the cellulose chain (Rabinovich et al., 1982; Stahlberg et al., 1988; Henriksson et al., 1999; Karlsson et al., 2002). EGs featuring CBMs can also hydrolyze cellulose chains at crystalline internal regions (Tilbeurgh et al., 1986; Gilkes et al., 1988; Tomme et al., 1988a; Wang et al., 2012f). Furthermore, EG cellulolysis generates new chain ends for cellobiohydrolase (CBH) activity (Wood and McCrae, 1972; Berghem and Pettersson, 1973; Henrissat et al., 1985). When an enzyme does not readily release a large molecular substrate and catalyzes multiple reactions before dissociating from the substrate, it is considered to be *processive*. Although processivity is typical of CBHs, some EGs also hydrolyze cellulose processively (Reverbel-Leroy et al., 1997; Irwin et al., 1998; Gilad et al., 2003; Cohen et al., 2005; Zverlov et al., 2005b; Zheng and Ding, 2013). EGs belong to GH families 5, 6, 7, 9, 12, 44, 45, 48, 51, 74, and 124 (Lombard et al., 2014; CAZy, 2015). All well-studied processive endocellulases are part of the GH9 family, which includes most plant cellulases, some animal cellulases, and many bacterial cellulases. Surprisingly, very few fungal cellulases are included within the GH9 family of endocellulases. Processive endoglucanases from GH9 family feature CBMs of the family 3c. These endoglucanase CBMs are positioned at the C terminus of the enzyme's catalytic domain (Sakon et al., 1997). They are required and responsible for processivity in this class of enzyme (Irwin et al., 1998; Gilad et al., 2003). A novel processive endoglucanase has been reported recently that belongs to the GH5 family (Cohen et al., 2005; Watson et al., 2009; Zheng and Ding, 2013).

EXs hydrolyze 1,4-β-D-glycosidic linkages in cello-oligosaccharides. Specifically, they cleave from the reducing or nonreducing ends of chains formed by EG activity. These exocellulases are processive and have their active sites in a *tunnel* (Rouvinen et al., 1990; Divne et al., 1994; Parsiegla et al., 1998). Several studies indicate that selected EXs, such as CBH, also cleave internal glycosidic bonds (Stahlberg et al., 1993; Armand et al., 1997; Boisset et al., 2000). There are two main groups of EXs: cellodextrinases and cellobiohydrolases (CBHs). Cellodextrinases (e.g., EC 3.2.1.74), also referred to as 1,4-β-D-glucan glucanohydrolases and exo-1,4-β-glucosidase, liberate D-glucose from cellodextrins and cellulose (Barras et al., 1969). Cellodextrinases belong to GH families 1, 3, 5, and 9 (Lombard et al., 2014; CAZy, 2015). CBHs (i.e., 1,4-β-D-glucan cellobiohydrolases) liberate D-cellobiose from cellulose chain ends produced by EGs and from crystalline cellulose (Kleman-Leyer et al., 1996; Igarashi et al., 2009; Liu et al., 2011), whereas CBHII (EC 3.2.1.91) also releases D-cellobiose from amorphous cellulose (Koivula et al., 1998) (Figure 3.1). CBHI (EC 3.2.1.176) works processively from the reducing end of cellulose, and CBHII works processively from the nonreducing end of cellulose (Fägerstam and Pettersson, 1980; Arai et al., 1989; Barr et al., 1996; Saharay et al., 2010). CBHIIs are grouped into GH families 5, 6, and 9. Most CBHIs belong to GH families 7 and 48 (Lombard et al., 2014; CAZy, 2015). Due to processivity and adsorption to insoluble cellulose substrates, CBH kinetics deviate from Michaelis–Menten model kinetics and exhibit fractal and *local jamming* effects (Xu and Ding, 2007; Igarashi et al., 2011; Kamat et al., 2013).

βGs (e.g., EC 3.2.1.21) hydrolyze the β-1,4-D-glycosidic bonds at the nonreducing ends of soluble cellooligosaccharides (e.g., cellodextrin) and cellobiose to release monomeric β-glucose (Freer, 1993). Unlike other carbohydrases, βGs generally lack distinct CBMs. Therefore, they do not have a modular architecture. Unlike the majority of biomass-degrading enzymes, βGs can be studied using traditional kinetic models (e.g., Michaelis–Menten) primarily due the fact that they act by binding to soluble substrate (Kempton and Withers, 1992; Chauve et al., 2010). βGs serve a critical role in multienzyme systems adding to synergistic effects by breaking down cellobiose, which produces glucose and minimizes the inhibition of cellulose activity by cellobiose (Berlin et al., 2005; Chir et al., 2011). Indeed, cellobiose is a strong inhibitor of CBH and EG (Holtzapple et al., 1990; Gusakov and Sinitsyn, 1992; Zhao et al., 2004; Andrić et al., 2010; Teugjas and Valjamae, 2013). Studies show that cellobiose inhibits cellulases 14 times more than glucose (Holtzapple et al., 1984).

The βGs belong to glycoside hydrolase (GH) families 1, 3, 5, 9, 30, and 116 (Lombard et al., 2014; CAZy, 2015). Different EGs possess different mechanisms of enzymatic activity (*inverting* for GH6, 9, 45, and 48 EGs; *retaining* for GH5, 7, and 12 EGs). Retaining will conserve the stereochemistry around the anomeric carbon atom from substrate to product.

Inverting will result in an axial to equatorial realignment of the anomeric carbon atom from substrate to product. The side activity of EGs on hemicellulose during a reaction may account for this ability to use more than one mechanism to catalyze reactions during the deconstruction of complex lignocellulose materials (Vlasenko et al., 2010). This may also serve in synergistic activity between processive and conventional EGs (Tuka et al., 1992; Qi et al., 2007).

3.1.1.2   Oxidoreductive cellulases

A nonhydrolytic factor that renders biomass less recalcitrant and more amenable to enzymatic action has been suspected for decades (Reese et al., 1950). An oxidative mechanism that could initiate cellulose degradation was suggested in the 1970s (Eriksson et al., 1974). It was later demonstrated that cellulose oxidases can disrupt cellulose structure via oxidation in a manner that results in an increased access to cellulose (Forsberg et al., 2011; Quinlan et al., 2011). Although cellulose oxidases occur in relatively low concentrations in natural systems, they play a central role in the cellulase systems of both aerobic fungi and bacteria that degrade cellulose (Harris et al., 2010; Forsberg et al., 2011). To date, cellulose oxidases have not been found in anaerobic cellulase complexes.

CDH, cellobiose quinone oxidoreductase (CBQOR), lactonase, glucose oxidase, and/or polysaccharide monooxygenases (PMOs) are commonly present in oxidoreductive enzyme systems (Figure 3.2).

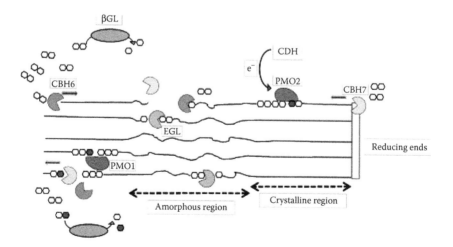

Figure 3.2 Schematic representation of the enzymatic degradation of cellulose, involving CBH, βG, EG, Type 1 PMOs (PMO1), Type 2 PMOs (PMO2), and CDH. (Adopted and modified from Dimarogona, M. et al., *Comput. Struct. Biotechnol. J.*, 2, 1–8, 2012.)

CDH (EC 1.1.99.18), also called cellobiose oxidoreductase (CBOR) or cellobiose oxidase (CBO), was originally named CBQOR (Westermark and Eriksson, 1974a, 1974b). One heme and one flavin adenine dinucleotide (FAD) serve as prosthetic groups for CDH (Ayers et al., 1978; Morpeth, 1985). As a streamline to the nomenclature, CBQOR is now used to refer to the catalytic active fragment of CDH, which lacks the heme group. CBQOR is produced by a posttranslational proteolytic cleavage and exhibits catalytic properties similar to CDH (Henriksson et al., 1991; Samejima and Eriksson, 1992; Wood and Wood, 1992; Raíces et al., 2002). Many wood-degrading fungi express CDH, and it is the only known example of a secreted flavocytochrome (Zamocky et al., 2006). CDH catalyzes the reducing end oxidation of cellobiose, cellodextrin, lactose, and maltodextrin and other oligosaccharides to the corresponding lactones using a host of electron acceptors, including quinones, phenoxyradicals, Fe^{3+}, Cu^{2+}, and triiodide (Henriksson et al., 2000). Spontaneous conversion or lactonase-mediated hydrolysis may then convert the lactones to aldonic acids (Brodie and Lipmann, 1955; Beeson et al., 2011).

Although CDH is reported to participate in the deconstruction of lignocellulose by generating hydroxyl radicals, its biological role is not fully understood. What is known is that CDH will reduce Fe^{3+} to Fe^{2+}, and in the presence of hydrogen peroxide, hydroxyl radicals are generated via Fenton's reaction ($H_2O_2 + Fe^{2+} \rightarrow Fe^{3+} + OH^{\bullet} + OH^-$) (Kremer and Wood, 1992a, 1992b). Furthermore, CDH will reduce cellulose inhibition by removing cellobiose (Ayers et al., 1978; Igarashi et al., 1998). Recent evidence also suggests the participation of CDH in the transfer of electrons to polysaccharide monooxygenases (PMOs), which results in the oxidative breakup of plant biomass (Phillips et al., 2011; Sygmund et al., 2012; Vu et al., 2014). CDHs belong to auxiliary activity family 3 (Lombard et al., 2014; CAZy, 2015).

Lactonase (EC 3.1.1.17), also called D-glucono-1,5-lactone lactonohydrolase, aldonolactonase, or gluconolactonase, catalyzes the hydrolysis of different types of hexose-1,5-lactones to their corresponding aldonic acids (Brodie and Lipmann, 1955; Beeson et al., 2011). Lactonase is found in commercial preparations of enzymes (e.g., Novozyme 188) from *Trichoderma reesei* and *Aspergillus niger* (Bruchmann et al., 1987). As a side effect, lactonase can facilitate cellulolysis as it removes lactones, which also inhibit cellulases (Bruchmann et al., 1987; Rouyi et al., 2014).

Glucose oxidase (EC 1.1.3.4), also known as notatin, is an oxidoreductase that mainly catalyzes the oxidation of glucose to hydrogen peroxide and D-glucono-δ-lactone, the latter of which hydrolyzes spontaneously to gluconic acid (Müller, 1928; Bentley and Neuberger, 1949). Glucose oxidase is also considered as an integral component of many cellulose systems because it acts to reduce the inhibition of cellulases by glucose (Holtzapple et al., 1984, 1990; Stutzenberger, 1986; Xiao et al., 2004; Andrić et al., 2010; Hsieh et al., 2014). Glucose oxidase belongs to the auxiliary activity family 3 (Lombard et al., 2014; CAZy, 2015).

Polysaccharide monooxygenases (PMOs) are copper-dependent metalloenzymes that oxidatively cleave glycosidic bonds at the surface of recalcitrant cellulose structures (Vaaje-Kolstad et al., 2010; Forsberg et al., 2011; Phillips et al., 2011; Quinlan et al., 2011). For catalytic efficiency, PMOs require a molecular oxygen and an electron donor, such as CDH (Phillips et al., 2011; Sygmund et al., 2012; Vu et al., 2014). PMOs introduce molecular oxygen to C–H bonds adjacent to the glycosidic linkage, which leads to the removal of the adjacent carbohydrate moiety (Phillips et al., 2011; Beeson et al., 2012). Based on substrate specificities and structures, PMOs are subdivided into *types*. Type 1 PMOs generate products that are oxidized at C-1. Type 2 PMOs generate products that are oxidized at the nonreducing end of C-4. Type 3 PMOs exhibit weaker specificity and release oxidized products from both reducing and nonreducing ends (Beeson et al., 2012; Li et al., 2012). PMOs belong to auxiliary activity families 9 and 10 (Lombard et al., 2014; CAZy, 2015).

3.1.1.3 Phosphorylases

Phosphorylases reduce cellobiose and cellodextrins to glucose using phosphates instead of water (Ayers, 1959; Alexander, 1961; Sheth and Alexander, 1967). Recall that cellobiose and cellodextrins are formed during the enzymatic degradation of cellulose by EGs and EXs. As phosphorolytic enzymes are located inside cells, organisms transport saccharides from the extracellular environment across the cell membrane. These become substrates for phosphorylase. For example, the thermophile *Clostridium thermocellum* employs ATP-driven transport mechanisms to uptake not only glucose but also cellobiose and cellodextrin (Strobel et al., 1995). These saccharides are then acted upon by phosphorylases to facilitate cellulose deconstruction. Although phosphorylases are not directly part of the cellulolytic pathway, phosphorolysis may accelerate the rate of overall cellulosic degradation when acting in concert with hydrolytic enzymes by removing inhibitory intermediate products such as cellobiose and cellodextrin (Holtzapple et al., 1990; Gusakov and Sinitsyn, 1992; Zhao et al., 2004; Andrić et al., 2010; Teugjas and Valjamae, 2013). Moreover, phosphorolysis is energetically advantageous. Phosphorolysis results in conservation of a portion of the energy from cleaved glycosyl bonds. Glucose-1-phosphate (G1P) leads to the formation of activated glucosyl molecules with the investment of only one ATP molecule for the uptake of a cellobiose or cellodextrin molecule. Each glucose molecule produced via

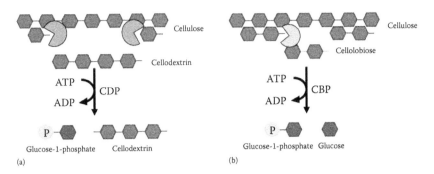

Figure 3.3 Chemical reactions catalyzed by (a) CDP: Cellodextrin + P_i ⇌ cellodextrin $(N - 1)$ + G1P and (b) CBP: Cellobiose + Pi ⇌ glucose + G1P.

hydrolytic cleavage would require two ATPs—one ATP for transport and another for activation (Goldberg, 1975; Strobel et al., 1995).

There are two general types of phosphorylases: cellodextrin phosphorylases (CDPs) and cellobiose phosphorylases (CBPs). CDPs (e.g., EC 2.4.1.49) phosphorylate cellodextrin released by endoglucanases to cellodextrin $(N - 1)$ and G1P (Sheth and Alexander, 1967). (Note: N is the number of glucose units in the chain). CBPs (e.g., EC 2.4.1.20) phosphorylate cellobiose into glucose and G1P as cellobiose is transported into the cell (Alexander, 1961) (Figure 3.3). Both enzymes belong to glycoside hydrolase family 94 (Lombard et al., 2014; CAZy, 2015).

3.1.2   Hemicellulases

Hemicellulases are either glycoside hydrolases or carbohydrate esterases (CEs) that catalyze the hydrolysis and deacetylation of hemicelluloses, respectively (Table 3.1). The mode of action of hemicellulases varies with the type of enzyme (Vocadlo and Davies, 2008; Biely, 2012). The heterogeneity and organization of hemicellulose require concerted and synergistic activity of multiple enzymes for complete degradation. Breakdown of the hemicellulose component in lignocellulosic feedstock exposes cellulose. This provides access for cellulases to primary substrates (e.g., cellulose) and results in the conversion of hemicellulose into usable saccharides. A principal way in which hemicellulose is deconstructed is through enzymatic action on xylan—a key component of hemicellulose (Timell, 1967). Enzymes which break down hemicellulose are generally called xylanases. However, there are several types of accessory enzymes that play an important role in degrading hemicellulose. Most of these accessory enzymes act on the exposed side chains of this heteropolymer to facilitate hemicellulose degradation.

Table 3.1 Summary of lignocellulosic enzyme classes

Enzyme	Abbreviation	Mode of action
Hydrolases		
α-L-arabinofuranosidase	AF	Nonreducing end of α-1,2-, α-1,3-, α-1,5-linked arabinofuranosyl groups from arabinans, arabinoxylans, and arabinogalactans
α-fucosidase	AFU	L-fucose residues from xyloglucan branches
α-galactosidase	AGL	Nonreducing end of α-linked D-nonreducing end galactose residues from xylan and galactomannans
α-D-glucuronidase	AgluA	Nonreducing end of α-1,2-linked 4-O-methyl-D-glucuronic acid residues from glucuronoxylans
Endo-1,5-α-arabinanase (arabinase)	AR	α(1→5) glycosidic bonds in arabinan
Arabinoxylan arabinofuranohydrolase	AXAH	Nonreducing end L-arabinofuranosyl groups from β-1,4-linked arabinoxylans
α-D-xyloside xylohydrolase or α-xylosidase	AXL	α-linked D-xylose residues from the xyloglucan backbone
β-glucosidase	βG	Nonreducing end of β-D-glucosyl residues from glucomannan and galactoglucomannan oligosaccharides
β-xylosidases	βX	Nonreducing ends of xylooligomers and xylobiose
Endogalactanase	EG	1,4-β-linked galactose residues in arabinogalactans
Endo-1,4-β-xylanase	EN	β-1,4-xylose linkages in heteroxylan backbone
Exo-β-1,4-xylanase	EXY	Reducing end of xylan backbone
β-galactosidase	LAC	Nonreducing end of β-linked D-galactose residues from xylan, xyloglucan, and galactoglucomannans

(Continued)

Table 3.1 **(Continued)** Summary of lignocellulosic enzyme classes

Enzyme	Abbreviation	Mode of action
Mannan endo-1,4-β-mannosidase, 1,4-β-D-mannan mannanohydrolase or endo-1,4-mannanase (β-mannanase)	MAN	β-1,4-linked bonds in mannan
β-mannosidase	MND	β-1,4-linked mannan oligosaccharides and mannobiose
Xyloglucan-β-1,4-endoglucanase or xyloglucanase (xyloglucan endohydrolase)	XGH	1,4-beta-D-glucosidic linkages in xyloglucan
Esterases		
Acetyl mannan esterase	AME	Acetyl groups from galactoglucomannan
Acetylxylan esterase	AXE	Acetyl esters in xylan and xylooligomers
Ferulic acid esterase or feruloyl esterase (cinnamoyl esterase hydrolases)	FAE	Monomeric or dimeric ferulic acid from xylans
Glucuronyl esterase	GE	4-*O*-methyl-D-glucuronic acid residues of glucuronoxylans
p-coumaric acid esterase or *p*-coumaroyl esterase	PAE	Monomeric and dimeric *p*-coumaric acid

3.1.2.1 Xylan-degrading enzymes

Xylan is naturally heterogeneous (Section 1.2.2 and Figures 1.1 and 1.4). Its hydrolysis requires the action of complex enzyme systems. Microbial enzymes act in a cooperative manner to convert xylan to its constituent simple sugars. The main enzymes involved are hydrolytic enzymes that hydrolyze β-1,4-xylosidic linkages. These enzymes are typically grouped into three classes: endo-1,4-β-xylanases (ENs), β-xylosidases (βXs), and exoxylanases (EXYs) (Figure 3.4).

ENs (e.g., EC 3.2.1.8) are glycosyl hydrolases that hydrolyze β-1,4-xylose linkages in the interior of the heteroxylan backbone to generate xylooligosaccharides. EN action on a substrate is determined by the chain length, the degree of branching, and/or the presence of specific substituents, such as arabinofuranosyl groups (Li et al., 2000; von Gal Milanezi et al., 2012). ENs also play a key role in lignin removal (Aracri and Vidal, 2011; Valenzuela et al., 2013). They deconstruct lignin-associated xylans,

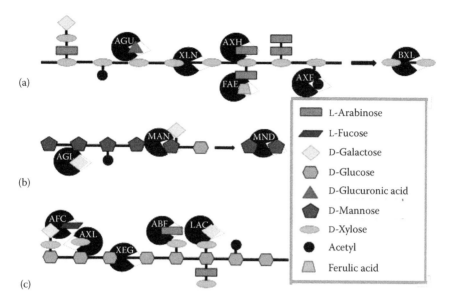

Figure 3.4 Schematic representation of (a) xylan, (b) galacto(gluco)man-nan, and (c) xyloglucan and mode of action of hemicellulolytic enzymes. AGU, alpha-glucuronidase; XLN, beta-1,4-endoxylanase; AXH, arabinoxylan arabinofuranohydrolase; FAE, feruloyl esterase; AXE, acetyl xylan; BXL, beta-1,4-galactosidase; MAN, beta-1,4-endomannanase; MND, beta-1,4-mannosidase; AFC, alpha-xylosidase; XEG, xyloglycan beta-1,4-endoglycanase; ABF, alpha-arabinofuranosidase; LAC, beta-1,4-galactosidase. (Adopted and modified from van den Brink, J. and de Vries, R.P., *Appl. Microbiol. Biotechnol.*, 91, 1477–1492, 2011.)

enhancing the accessibility and extractability of lignin (Roncero et al., 2005). ENs are categorized into families 5, 8, 10, 11, 30, 43, and 51 of the glycoside hydrolases based on amino acid sequence similarities (Lombard et al., 2014). GH10 and GH11 ENs differ in substrate specificity. GH10 ENs are capable of cleaving glycosidic linkages in the xylan main chain adjacent to substituents in which there is a functional group (or other atom) substitute in place of a hydrogen atom on a hydrocarbon. GH11 ENs preferentially cleave unsubstituted regions. As a result, GH10 ENs release products that are shorter than the products of GH11 EN activity (Biely et al., 1997; Ustinov et al., 2008).

βXs (e.g., EC 3.2.1.37) release monomeric xylose from the nonreducing ends of xylooligomers and xylobiose produced by EN action on xylan. βXs have molecular weights >100 kDa and typically consist of two or more subunits (Matsuo and Yasui, 1984; Hebraud and Fevre, 1990; Eneyskaya et al., 2003, 2007). They have catalytic cores similar to those found in GH families 1, 3, 30, 39, 43, 51, 52, 54, 116, and 120 (Lombard et al., 2014; CAZy, 2015). Generally, βX activity on xylooligosaccharides decreases rapidly with increasing chain length (Van Doorslaer et al., 1985; Rasmussen et al., 2006). Many βXs exhibit α-ʟ-arabinofuranosidase activity, and some βXs reportedly have β-glucosidase activity (Rodionova et al., 1983; Uziie et al., 1985; Xiong et al., 2007; Watanabe et al., 2015). Notably, most βXs are susceptible to xylose inhibition, which can significantly affect enzymatic efficiency under processing conditions (Dekker, 1983; Poutanen, 1988; Herrmann et al., 1997; Saha, 2003b; Fujii et al., 2011; Kirikyali and Connerton, 2014). By the breakdown of xylobiose, βXs relieve EN end product inhibition (Sunna and Antranikian, 1997; Williams et al., 2000).

The EXYs (e.g., EC 3.2.1.156) are the most recently characterized xylandegrading enzymes. Only a few exoxylanases have been reported, and information on their catalytic properties is limited. It is known that EXYs act on the xylan backbone from the reducing end to release xylose and to produce short xylooligomers (Ganju et al., 1989; Kubata et al., 1994, 1995; Usui et al., 1999; Honda and Kitaoka, 2004; Fushinobu et al., 2005; Tenkanen et al., 2013; Juturu et al., 2014).

EXYs differ from βXs in that the former does not break down on xylobiose (Kubata et al., 1994, 1995). EXYs can increase the rate of hydrolysis of xylan, because endoxylanase would increase the ends available on the xylan backbone to the exoxylanase (Gasparic et al., 1995; Juturu et al., 2014). EXYs belong to glycoside hydrolase family 8 (Lombard et al., 2014; CAZy, 2015).

3.1.2.2 *Accessory enzymes*

Accessory enzymes either degrade the side chains of xylans (debranching enzymes) or act on the backbone chains of different kinds of hemicelluloses (backbone-degrading enzymes). They are more generally either hydrolases or esterases.

A wide variety of xylanolytic accessory enzymes can enhance hemicellulose deconstruction, including α-ʟ-arabinofuranosidases (AFs); arabinoxylan arabinofuranohydrolases (AXAHs); endo-1,5-α arabinanases cumulatively known as arabinases (ARs); xylan α-ᴅ-glucuronidases or xylan α-1,2-ᴅ-glucuronidase (AgluAs); mannan endo-1,4-β-mannosidase, 4-β-ᴅ-mannan mannanohydrolase, endo-1,4-mannanase or, simply, the β-mannanases (MANs); β-mannosidases (MNDs); α-galactosidases (AGLs) and β-galactosidases (LACs), or, simply, the galactosidases; β-glucosidases (βGs); endo-β-1,4-galactanases (EGs); xyloglucan-β-1,4-endoglucanases or xyloglucanase, cumulatively known as the xyloglucanendohydrolases (XGHs); α-ᴅ-xyloside xylohydrolase or, simply, α-xylosidases (AXLs); α-fucosidases (AFUs); acetylxylan esterases (AXEs); ferulic acid esterases or feruloyl esterases, also known as the cinnamoyl esterase hydrolyses (FAEs); *p*-coumaric acid esterases or *p*-coumaroyl esterases (PAEs); glucuronyl esterases (GEs); and acetyl mannan esterases (AMEs) (Table 3.1).

The AF (e.g., EC 3.2.1.55) catalytic domains belong to GH2, 3, 43, 51, 54, and 62 families of hydrolases (Lombard et al., 2014; CAZy, 2015). These enzymes cleave the nonreducing end α-1,2-, α-1,3-, and α-1,5-linked L-arabinofurano-syl groups in hemicellulose, such as arabinoxylan or arabinogalactan (Saha and Bothast, 1998; Verbruggen et al., 1998a; Ahmed et al., 2013). This mode of action is effective in hydrolyzing hemicellulose side chains and in disrupting structures anchored by α-glycosidic bonds. AXAHs (e.g., EC 3.2.1.55) are essentially AFs from the GH51 family that specifically remove the terminal nonreducing arabinofuranosyl from the arabinoxylan 1,4-β-xylan backbone (Kormelink et al., 1991; Ferre et al., 2000; Lee et al., 2001).

Both AFs and AXAHs facilitate the disruption of lignin–carbohydrate binding at locations where arabinose residues are involved in lignin–hemicellulose ether bonds (Sun et al., 2005). AFs that exhibit β-xylosidase or xylanase catalytic activity have also been reported in the literature (Utt et al., 1991; Matte and Forsberg, 1992; Mai et al., 2000; Lee et al., 2003). Similar to AFs, ARs (e.g., EC 3.2.1.99) with catalytic domains belonging to the GH43 family also cleave internal α(1→5) glycosidic bonds in arabinan (Hong et al., 2009; Lombard et al., 2014; Shi et al., 2014).

AgluAs (e.g., EC 3.2.1.131), of the GH67 and GH115 families of hydro-lases, are reported to cleave α-1,2-glycosidic bonds of the 4-O-methyl-D-glucuronic/D-glucuronic acid residues from the terminal nonreducing xyloses of glucuronoxylooligosaccharide or polymeric glucuronoxylan (Tenkanen and Siika-aho, 2000; Nurizzo et al., 2002; Ryabova et al., 2009; Lee et al., 2012a; Lombard et al., 2014; Rogowski et al., 2014; CAZy, 2015).

MANs (e.g., EC 3.2.1.78) disrupt internal β-1,4-linkages in backbone structure of mannan polymers producing new chain ends while also releasing short β-1,4-mannooligosaccharides (Mandels and Reese, 1965; Stålbrand, 1993; Katrolia et al., 2013). MANs are in GH5, GH26, and GH113 families (Lombard et al., 2014; CAZy, 2015). MNDs (e.g., EC 3.2.1.25) cleave β-1,4-linked mannooligosaccharides and mannobiose from the break-down products of the endomannases (e.g., MANs) producing mannose by acting on these products at the nonreducing terminal ends (Gübitz et al., 1996; Andreotti et al., 2005; Zhang et al., 2009). MNDs are GH1, 2, and 5 enzymes (Lombard et al., 2014; CAZy, 2015).

Galactosidases are glycoside hydrolases that catalyze the hydrolysis of galactosides into monosaccharides. There are two types of galactosi-dases. AGLs (e.g., EC 3.2.1.22), which belong to GH4, 27, 31, 36, 57, 97, and 110 families, release α-linked D-galactose residues from hemicellulose (i.e., xylan or galactomannan) by acting at the nonreducing terminal ends (Ademark et al., 2001; Lombard et al., 2014; CAZy, 2015). LACs (e.g., EC 3.2.1.23), belong to GH1, 2, 35, 42, and 59 families (Lombard et al., 2014; CAZy, 2015). LACs hydrolyze the nonreducing ends of β-D-galactose resi-dues from hemicellulose (i.e., xylan, xyloglucan, or galactoglucomannan) (Sims et al., 1997). The βGs (EC 3.2.1.21) are also exotype enzymes that remove the 1,4-β-D-glucopyranose units from the nonreducing ends of oligosaccharides from the breakdown of glucomannan and galactogluco-mannan by MAN (Moreira and Filho, 2008).

EGs (e.g., EC 3.2.1.89) belong to GH53 family and hydrolyze 1,4-β-linked galactose residues in arabinogalactans (Lombard et al., 2014; CAZy, 2015). XGHs (e.g., EC 3.2.1.151) hydrolyze xyloglucans into oligoxy-loglucans and belong to GH5, 9, 12, 16, and 44 families (Lombard et al., 2014; CAZy, 2015). AXLs (e.g., EC 3.2.1.177) release D-xylose residues with α-linkages from the nonreducing terminal of xyloglucooligosaccharide (Ariza et al., 2011; Larsbrink et al., 2011). They belong to GH31 family (Lombard et al., 2014; CAZy, 2015). AFUs (EC 3.2.1.51), belonging to GH29 and 95 families, release L-fucose residues from xyloglucan branches (Léonard et al., 2008; Lombard et al., 2014; CAZy, 2015). These represent the major hydrolases.

In addition to hydrolases acting on glycosidic linkages in hemicellulose, CEs catalyze O- or N-deacylation of substituted saccharides. Only CEs that act on sugars serving as acids within hemicellulose structures, such as in acetylated xylan, are considered here. CEs are grouped into several differ-ent enzyme classes. AXEs belong to enzyme families CE1, 2, 3, 4, 5, 6, 7, and 12 (Lombard et al., 2014; CAZy, 2015). AXEs (e.g., EC 3.1.1.72) hydrolyze acetyl ester bonds at the C-2, C-3, and C-4 positions of xylose in both xylan and xylo-oligomers (thus removing O-acetyl groups) (Biely, 2012). Ferulic acid esterases (FAEs) belong to CE1 family. FAEs (e.g., EC 3.1.1.73) hydrolyze hydroxycin-namoyl ester bonds liberating hydroxycinnamic acids such as monomeric or dimeric ferulic acid (Lombard et al., 2014; CAZy, 2015). FAE action can target either the O-2 or O-5 on α-L-arabinoses in xylan. Cinnamoyl substitu-tion can occur via hydroxylation or methoxylation.

Both the nature of the cinnamoyl substitution and the type of hemicel-lulose linkage formed (e.g., arabinose versus galactose ester bonding with xylan or pectin) will determine the specificity of a given FAE (Benoit et al., 2008). The p-coumaric acid-type esterases (PAEs) are classified within the CE1 family (Lombard et al., 2014; CAZy, 2015).

PAEs hydrolyze ester linkages between arabinose side chain residues of phenolic acids, including monomeric and dimeric p-coumaric acid (PA) (Borneman et al., 1991). Glucuronoyl esterases (GEs) are found within the CE15 family (Lombard et al., 2014; CAZy, 2015). GEs hydrolyze methyl ester bonds between 4-O-methyl-D-glucuronic acid residues of glucuro-noxylans and aromatic alcohols, which are found in lignin (Špániková and Biely, 2006; Ďuranová et al., 2009). Another acetyl esterase that acts on mannan (AME) is found within the CE16 family of enzymes (Lombard et al., 2014; CAZy, 2015). AMEs (e.g., EC 3.1.1.6) release acetyl groups from mannan containing units, such as galactoglucomannan (Shallom and Shoham, 2003).

3.1.3 Ligninolytic enzymes

Many *white-rot* fungi are capable of acting on the lignin component of ligno-cellulosic biomass. Such fungi break down lignin in plant cell walls, releasing CO_2 and H_2O and carbohydrates, which are used as a source of carbon and energy (Fackler et al., 2006; Arora and Sharma, 2009). Ligninolysis by *brown-rot* fungi is also reported; however, there are few reports on lignin breakdown by soil bacteria (Crawford et al., 1983; Mercer et al., 1996; Kirby, 2006; Bugg et al., 2011). In general, there appear to be two major families of oxidative enzymes (oxidoreductases) involved in ligninolysis: peroxidases and laccases. The peroxidases are further subdivided into lignin peroxidases (LiPs), manganese-dependent peroxidases (MnPs), versatile peroxidases (VPs), and dye-decolorizing peroxidases (DyPs). Lac enzymes catalyze a single-electron oxidation of lignin. The transfer of one electron in each step from aromatic lignin moieties (with low reduction potential) to the high redox potential active sites in the enzyme generates highly reactive nonspecific free radicals (e.g., reactive oxygen species). These nonspecific free radicals facilitate lignin depolymerization via nonenzymatic reactions (Harvey et al., 1985; Schoemaker et al., 1985; Hammel et al., 2002). Several accessory enzymes, including oxidases, are also involved in the breakdown of lignin by increasing the ligninolytic activity of principal enzymes.

Notably, many of the enzymes involved in lignin deconstruction in woody biomass cannot penetrate the compact structure of woody tissues (Srebotnik et al., 1988; Flournoy et al., 1993; Blanchette et al., 1997). However, the enzymes can act on the surface of the cell wall to produce low molecular mass agents (Evans et al., 1994), which can diffuse through the cell wall and initiate wood decay. On active decay from within woody biomass, lignin-degrading enzymes may gain access to substrate (Galkin et al., 1998).

3.1.3.1 Peroxidases

Extracellular heme peroxidases, which are part of the auxiliary activity family 2, exhibit high potency in oxidative degradation of lignin and require extracellular H_2O_2 as an electron acceptor (Lombard et al., 2014; CAZy, 2015). On interaction with H_2O_2, these enzymes form highly reactive oxo-ferryl species (e.g., Fe^{5+}- or Fe^{4+}-oxo species), which serve as intermediates in catalytic reactions. These oxoferryl species remove electrons from lignin, causing oxidation or radicalization and subsequent ligninolytic action.

Note that with MnPs, an oxoferryl oxidizes Mn^{2+} to Mn^{3+}, which drives lignin oxidation (Wong, 2009) (Figure 3.5). LiPs and MnPs were discovered in *Phanerochaete chrysosporium* (ca. 1980) and exhibit high redox potential (Tien and Kirk, 1983; Kuwahara et al., 1984; Millis et al., 1989). Both LiPs and MnPs as well as VPs and DyPs (described later) are commonly found in cellulose degradation systems.

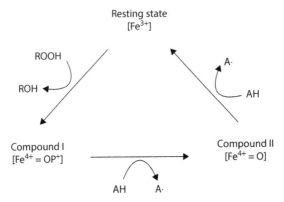

Figure 3.5 General catalytic cycle of heme-containing peroxidases: First, the resting state (Fe^{3+}) is involved in two-electron oxidation with H_2O_2 to form a Compound I oxoferryl intermediate ($Fe^{4+} = OP^+$). Then, Compound I oxidizes electron donor substrates (AH) by one-electron oxidation, yielding Compound II ($Fe^{4+} = O$) and a substrate cation radical (A·). The last step implies another oxidation of substrate by Compound II subtracting one electron and consequently, returning the enzyme to the resting state (Veitch, 2004). (Adopted from Furukawa, T. et al., *Front. Biol.*, 9, 448–471, 2014.)

LiP (e.g., EC 1.11.1.14) was first discovered in *P. chrysosporium*. LiPs catalyze the depolymerization of lignin through H_2O_2-dependent oxidation (Tien and Kirk, 1983, 1984). There are typically several isoenzymes of LiP encoded by different genes within a single fungus (Glumoff et al., 1990; Johansson et al., 1993). LiPs are strong oxidants with redox potentials that typically exceed 1 V; thus, LiPs are strong oxidizing agents when compared to other peroxidases (Ward et al., 2003). Underlying this potential is a porphyrin iron ring moiety that endows LiPs with a higher electron deficiency than that found in classical peroxidases (Millis et al., 1989). LiPs are considered to be a highly effective (and versatile) class of lignin-degrading peroxidases. LiPs oxidize both phenolic and nonphenolic compounds (Mester et al., 2001; Ward et al., 2001). LiP-catalyzed reactions include (1) cleavage of C–C bonds; (2) cleavage of ether (C–O–C) bonds in nonphenolic aromatic substrates; (3) hydroxylation of benzylic methylene groups; (4) oxidation of benzyl alcohols to aldehydes or ketones; (5) phenol oxidation; and (6) aromatic cleavage of nonphenolic lignin model compounds (Tien and Kirk, 1984; Hammel et al., 1985; Leisola et al., 1985; Renganathan et al., 1985, 1986; Umezawa et al., 1986). Because LiP is physically too large to enter a plant cell, LiP-mediated catalysis works within exposed regions of the lumen.

At the molecular level, MnPs (e.g., EC 1.11.1.13) are similar to LiPs. MnPs are extracellular heme enzymes that use manganese as a cofactor (Glenn and Gold, 1985; Paszczyński et al., 1986). MnPs were also discovered in the white-rot fungus *P. chrysosporium* (Kuwahara et al., 1984). MnP functions by oxidizing Mn^{2+} to Mn^{3+} in the presence of H_2O_2 (Glenn et al., 1986). This results in the formation of a Mn^{3+} oxalate complex. (However, note that the Mn^{2+} ion must first be chelated by organic acid chelators to stabilize the Mn^{3+} product). This also produces diffusible oxidizing chelates (Glenn and Gold, 1985; Glenn et al., 1986; Perez and Jeffries, 1992). Although Mn^{3+} is a strong oxidant that can leave the active center and oxidize phenolic compounds, unlike the LiPs, MnP-mediated catalysis cannot attack non-phenolic units of lignin (Popp and Kirk, 1991). However, resulting phenoxy radicals undergo a variety of reactions ultimately leading to lignin depolymerization (Tuor et al., 1992). Through the peroxidation of unsaturated lipids in the presence of Mn^{2+}, it has been reported that MnPs can oxidize model nonphenolic lignin compounds (Jensen et al., 1996; Kapich et al., 2005, 2010). Thus, it has been proposed that expression of both MnP and laccase in white-rot fungi, which do not produce LiP, may enable MnPs to cleave nonphenolic lignin substrates (Reddy et al., 2003).

VP (EC 1.11.1.16) from a white fungus, *Pleurotus eryngii*, was reported as a novel peroxidase possessing both MnP and LiP activity. VP oxidizes both phenolic and nonphenolic aromatic compounds, including veratryl alcohol and *p*-dimethoxybenzene. VP oxidizes Mn^{2+} similar to MnPs (Martinez et al., 1996a, 1996b; Camarero et al., 1999; Ruiz-Duenas et al., 1999). Similar to LiPs, VP has a high redox potential on nonphenolic compounds (Camarero et al., 1999). VP is a heme-containing ligninolytic peroxidase with a hybrid molecular structure consisting of different active sites that mediate oxidation (Pérez-Boada et al., 2005; Ruiz-Duenas et al., 2009).

If Mn^{2+} is available for the reaction, VP can oxidize hydroquinone without exogenous H_2O_2. As ligninolytic enzymes cannot typically penetrate unmodified or untreated wood cell walls, chemical oxidation of hydroquinones promoted by Mn^{2+} may be important for the initial steps of wood biodegradation (Gomez-Toribio et al., 2001).

DyPs (e.g., EC 1.11.1.19) were first described from studies of fungi (Kim and Shoda, 1999). Similar to MnPs and VP, DyPs are heme-containing peroxidases; however, there is a little similarity in primary sequence and structure between DyP and other plant, bacterial, and fungal peroxidases. Furthermore, DyPs appear to perform better than other peroxidases in lower pH environments (Sugano et al., 1999, 2007; Sugano, 2009). DyPs have broad substrate specificity.

Not only are DyPs able to oxidize all of the typical peroxidase substrates, they can also oxidize high-redox potential dyes (i.e., anthraquinones), which other peroxidases do not (Kim et al., 1995; Kim and Shoda, 1999; Sugano et al., 2000; Liers et al., 2010; Santos et al., 2014). It is clear that DyPs exhibit ligninolytic activity *in vitro*; however, their role in natural systems is less clear. Some evidence suggests a role for DyPs in lignin degradation in the environment (Sugano, 2009; Adav et al., 2010;

Ahmad et al., 2011; Salvachúa et al., 2013). However, more details are required to determine the extent to which this may be the case.

3.1.3.2 Laccases

Laccases (Lacs; e.g., EC 1.10.3.2) are glycosylated multi-copper phenoloxi-dases found in plants, fungi, and bacteria (Dwivedi et al., 2011). They are classified in the auxiliary activity family 1 (Lombard et al., 2014; CAZy, 2015). Lacs are found ubiquitously across wood-degrading fungi (Baldrian, 2006). Lacs do not require manganese ions or hydrogen perox-ide for catalysis. They catalyze a single-electron oxidation of substrates via a concomitant four electron reduction of O_2 to H_2O (Solomon et al., 1996, 2001; Messerschmidt, 1997). Laccases can directly oxidize pheno-lic components in lignin. Direct oxidation of phenolic lignin generates phenoxy-free-radicals, which ultimately lead to polymer cleavage (Kawai et al., 1988). Lacs do not act on substrates with low redox potentials since they have low redox potentials (≤ 0.8 V) (Reinhammar, 1972; Schneider et al., 1999; Johnson et al., 2003; Uzan et al., 2010). For example, nonpheno-lics, which have a redox potential above 1.3 V (Zweig et al., 1964), are not directly oxidized by Lacs. Lacs require the assistance of suitable media-tors to break down nonphenolic components. On oxidation by Lacs, these low molecular weight compounds are converted to radicals and act as redox mediators that, in turn, oxidize other compounds that are not directly Lac substrates (Bourbonnais and Paice, 1990; Barreca et al., 2003; Cho et al., 2008). Thus, if Lacs are used directly as a biomass pretreatment (i.e., for bioethanol production), the addition of exogenous mediators, such as 2,29-azinobis(3-ethylbenzthiazoline-6-sulfonate) (Bourbonnais and Paice, 1990), may not be necessary, because natural mediators will be generated. For example, Lac activity on phenolic lignin units will result in the release of phenoxy radicals.

These radicals will then serve as natural mediators and oxidize more recalcitrant nonphenolic lignin moieties (d'Acunzo et al., 2006; Nousiainen et al., 2009). It has also been reported that Lacs exhibit demethylating activity on lignin subunits (Trojanowski et al., 1966; Harkin and Obst, 1974; Ishihara and Miyaxaki, 1976; Leonowicz et al., 1979; Ander et al., 1983; Malarczyk et al., 2009) and lignin preparations (Ishihara and Miyaxaki, 1974; Ander and Eriksson, 1985; Crestini and Argyropoulos, 1998; Ibrahim et al., 2011). During demethylation, lacs act on aryl-*O*-alkyl C–O bonds.

3.1.3.3 Accessory enzymes and mediators

There are some accessory enzymes and nonenzymatic components known as mediators that increase the ligninolytic activity of primary enzymes. Some of these mediators are involved in the production of hydrogen perox-ide, which is required by peroxidases. However, others catalyze the reduc-tion of phenolic products derived from lignin degradation preventing repolymerization. Many of these enzymes are oxidases and reductases. Glyoxal oxidase (GLOX) is a copper radical enzyme (i.e., EC 1.2.3.5) originally isolated from the white-rot fungus *P. chrysosporium* (Kersten and Kirk, 1987; Kersten, 1990; Takano et al., 2010). Aryl alcohol oxidase (AAO), which is also known as veratryl alcohol oxidase (VAO) (EC 1.1.3.7), was found in *P. eryngii* (Guillén et al., 1990; Hernandez-Ortega et al., 2012). Other accessory oxidases include pyranose 2-oxidase or glucose 2-oxidase (EC 1.1.3.10) (Ruelius et al., 1968; Janssen and Ruelius, 1968a; Volc et al., 1978; Daniel et al., 1994); glucose oxidase (or glucose 1-oxidase) (EC 1.1.3.4) (Muller, 1928, 1936; Franke and Lorenz, 1937; Franke and Deffner, 1939; Kelley and Reddy, 1986); and alcohol oxidase (AOX) or methanol oxidase (EC 1.1.3.13) (Janssen et al., 1965; Janssen and Ruelius, 1968b; Suye, 1997). In addition, fungi produce reductases such as aryl-alcohol dehydrogenases (AAD) (EC 1.1.1.91) (Muheim et al., 1991; Gutierrez et al., 1994); quinone reductases (QR) (EC 1.6.5.5) (Guillen et al., 1997; Bazzi, 2001); and CDH also known as CBOR, CBO, or CBQOR (EC 1.1.99.18) (Westermark and Eriksson, 1974a, 1974b; Temp and Eggert, 1999). Each of these mediators enhances the activity of other lignocellulose deconstruction enzymes and contributes to the synergistic effects of multienzyme systems on substrate degradation. In both natural systems and industrial enzyme cocktails, the use of accessory enzymes and mediators can significantly enhance lignocellulose substrate breakdown.

3.1.4 Pectinolytic enzymes (Pectinases)

Another class of enzymes is the pectinases. Pectinase catalyzes the cleavage of pectic compounds (e.g., pectin). Depending on the cleavage sites utilized, pectinases are categorized into one of the three groups: esterases, lyases, and hydrolases (Sharma et al., 2013a).

3.1.4.1 Main pectinases

The most studied pectinolytic enzymes are homogalacturonan-degrading enzymes: polygalacturonases (PGs) or pectin depolymerase; polymethylgalacturonases (PMG); lyases or transeliminases; and pectinesterases (PEs), which are also called pectin methylesterases (PMEs).

PGs catalyze the hydrolytic cleavage of α (1,4)-glycosidic linkages in polygalacturonic acid chains to form D-galacturonate. They belong to glycosyl hydrolase family 28 (Markovic and Janecek, 2001; Lombard et al., 2014; CAZy, 2015). PGs are subdivided into endopolygalacturonases (PGA) and exopolygalacturonases (XPG). PGA (e.g., EC 3.2.1.15) acts on pectic acid (polygalacturonic acid) to produce several types of Gal A oligosaccharides. PGA disrupts internal α-1,4-D-glycosidic linkages between nonmethylated (or low methyl esterified) acid residues in pectic acid. PGA primarily acts on glycosidic bonds adjacent to galacturonic acids with free

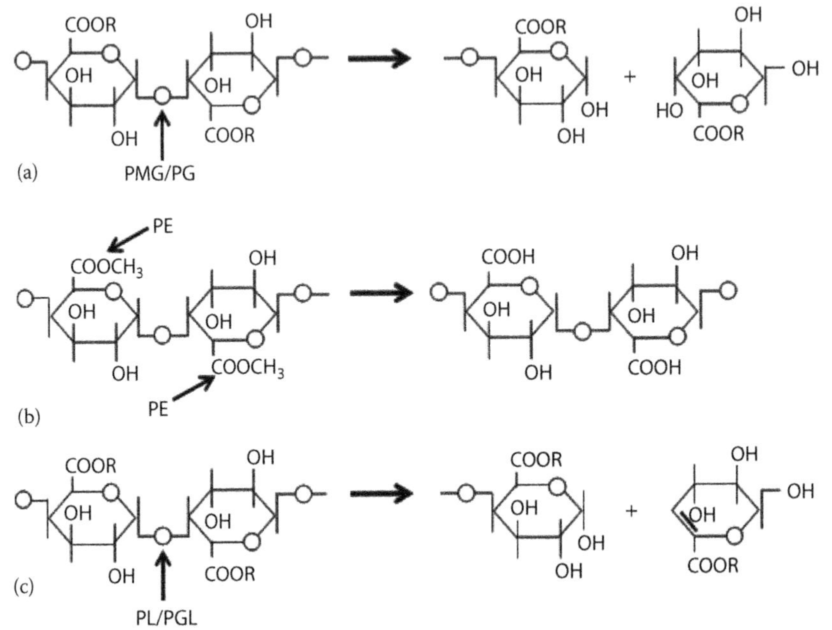

Figure 3.6 Mode of action of pectinases: (a) R=H for PG and CH_3 for PMG, (b) PE, and (c) R=H for PGL and CH_3 for PL. The place where the pectinolytic enzymes react with pectin are shown by the arrows. (Adopted from Pedrolli, D.B. et al., *Open Biotechnol. J.*, 3, 9–18, 2009.)

carboxyl groups (Yuan and Boa, 1979; Mohamed et al., 2006). XPG (e.g., EC 3.2.1.67) catalyzes the hydrolysis of α-1,4-glycosidic linkages of the non-reducing end HG chains releasing monogalacturonate. XPG requires a nonesterified GalpA unit at selected subsites (−2, −1, and +1) for optimal activity (Kester et al., 1999b). XPG is tolerant of xylose substitution and will act to remove Gal A–Xyl dimers. Therefore, XGA is also a viable XPG substrate (Beldman et al., 1996; Kester et al., 1999a) (Figure 3.6a).

PMG hydrolyzes α-1,4-glycosidic linkages of the pectin backbone and works efficiently on highly esterified pectin, forming 6-methyl-D-galacturonate bonds in the resulting product (Seegmiller and Jansen, 1952) (Figure 3.6a).

Lyase performs a transeliminative reaction on α-1,4 glycosidic bonds of polygalacturonic acid polymers to form Δ-4,5 unsaturated C–C bonds at the nonreducing ends of the cleaved pectin polysaccharides (Albersheim et al., 1960; Moran et al., 1968). There are two types of lyases: pectate lyase or pectate transeliminase (PGL) and pectin lyase or pectin transeliminase (PL). PGL acts on pectin to cleave glycosidic bonds generating unsaturated oligogalacturonates or digalacturonates. PGLs are Ca^{2+} dependent and act specifically on nonesterified pectin (pectate) (Starr and Moran,

1962; Pickersgill et al., 1994; Mayans et al., 1997; Seyedarabi et al., 2010). The enzyme is grouped into five of the polysaccharide lyase families: 1, 2, 3, 9, and 10 (Lombard et al., 2014; CAZy, 2015). PGLs are divided into two types: endo-PGL and exo-PGL. The endo-PGLs (e.g., EC4.2.2.2) act on substrates at random internal sites within the chain. The exo-PGLs (e.g., EC 4.2.2.9) act on the reducing end to catalyze substrate cleavage. Unlike PGLs, PLs (e.g., EC 4.2.2.10) are not Ca^{2+} dependent. PLs catalyze the random cleavage of highly esterified pectin and produce unsaturated methyloligogalacturonates (Figure 3.6c) (Albersheim et al., 1960; Edstrom and Phaff, 1963; Delgado et al., 1992; Vitali et al., 1998). PLs belong to polysaccharide lyase family 1 (Lombard et al., 2014; CAZy, 2015).

PE or PME (EC 3.1.1.11) catalyzes the deesterification of methyl ester linkages. Specifically, pectin (methyl) esterases catalyze the hydrolysis of ester bonds to pectate and methanol, thus removing the methoxyl group at O-6. There is preference for methyl ester groups in galacturonate units adjacent to nonesterified galacturonate units (Solms and Deuel, 1955; Fries et al., 2007). In natural systems, this enzyme is active prior to PG and PGL activity, the latter of which requires nonesterified substrates. After PE/PME activity, PG and lyase will act on the resulting pectin. PE is a part of carbohydrate esterase family 8 (Lombard et al., 2014; CAZy, 2015) (Figure 3.6b).

3.1.4.2   Other pectinases

Several other types of pectinases are not well studied. These include pectin acetyl esterases (PAEs); rhamnogalacturonase (RGase) also known as rhamnogalacturonan hydrolases (RGHs); rhamnogalacturonan rhamnohydrolases (RGRHs); rhamnogalacturonan galacturonohydrolases (RGGHs); rhamnogalacturonan endolyases (RGLs); rhamnogalacturonan acetylesterases (RGAs); xylogalacturonan hydrolases (XGHs); and other accessory enzymes (Figure 3.7). These enzymes are either backbone-degrading enzymes or debranching enzymes. PAE hydrolyzes the acetyl ester group of homogalacturonan (HG) and rhamnogalacturonan (RGI) forming pectic acid and acetate (Williamson et al., 1990; Williamson, 1991; Shevchik and Hugouvieux-Cotte-Pattat, 1997; Bolvig et al., 2003; Bonnin et al., 2008). It belongs to carbohydrate esterase families 12 and 13 (Lombard et al., 2014; CAZy, 2015). RGase/RGH (EC 3.2.1.171) is an endoacting enzyme capable of randomly hydrolyzing the α-D-1,4-GalpA-α-L-1,2-Rhap linkage in the RGI backbone, thereby producing oligogalacturonates. RGase/RGH is not efficient on the substrate that includes acetyl esterification of the RGI backbone (Schols et al., 1990; Kofod et al., 1994). RGase/RGH is grouped into glycosyl hydrolase family 28 (Lombard et al., 2014; CAZy, 2015). RGRH (EC 3.2.1.174) is an exoacting pectinase that catalyzes the hydrolytic cleavage of the rhamnogalacturonan chain of RGI at the nonreducing ends yielding rhamnose (Mutter et al., 1994).

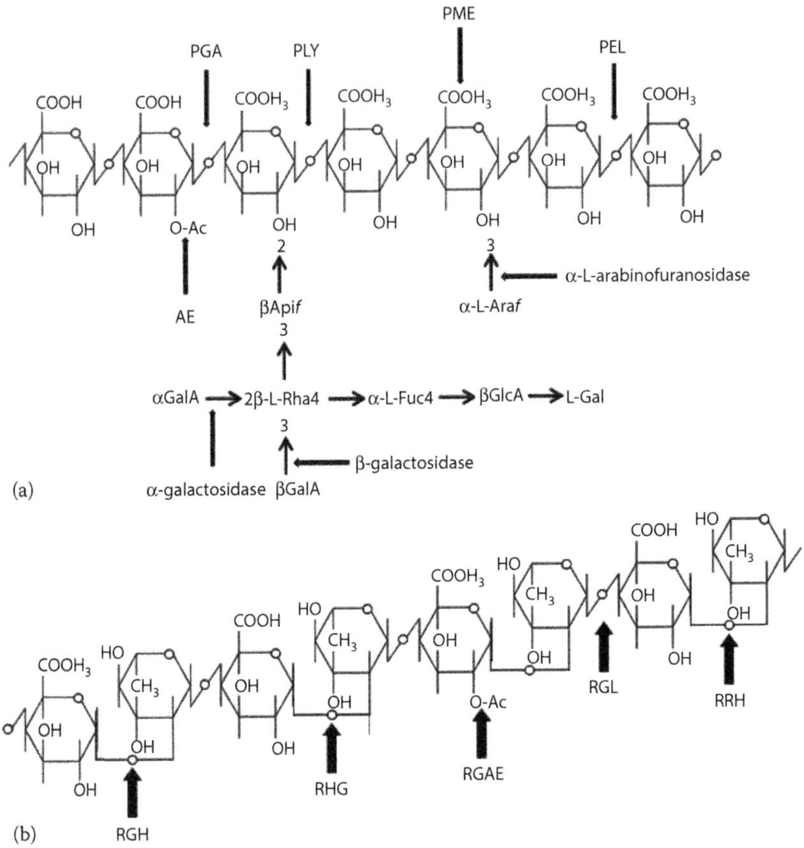

Figure 3.7 (a,b) schematic representation of HG and lateral chains of RG-II, and mode of action of pectinolytic enzymes involved in their degradation. AE, acetylesterases; Apif, beta-d-apiofuranosyl; Araf, alpha-l-arabinofuranosyl; PLY, pectate lyase; RGH/RHG, endorhamnogalacturonan hydrolases; RRH, rhamnogalacturonan rhamnohydrolase; RGL, rhamnogalacturonan lyase; PEL, pectin lyase. (Adopted and modified from Lara-Marquez, A., Zavala-Paramo, M.G., Lopez-Romero, E., Camacho, H.C., Biotechnological potential of pectinolytic complexes of fungi, *Biotechnol. Lett.*, 33, 859–868, 2001).

RGRH belongs to glycosyl hydrolase family 28 (Lombard et al., 2014; CAZy, 2015). RGGH (EC 3.2.1.173) is also an exoacting pectinase that catalyzes the hydrolytic cleavage of the rhamnogalacturonan chain of RGI at the nonreducing end; however, it produces monogalacturonate, a GalA moiety (Mutter et al., 1998a). RGGH belongs to family 28 of the glycosyl hydrolase family (Lombard et al., 2014; CAZy, 2015). RGL (EC 4.2.2.23) catalyzes the random transelimination (β-elimination) of the RGI α-L-1,2-Rhap-α-D-1,4-GalpA backbone, yielding unsaturated galacturonate at the nonreducing end and rhamnose at the reducing end (Kofod et al., 1994; Mutter et al., 1996). RGL activity may be hindered by the presence of the acetyl groups in the RGI backbone (Kofod et al., 1994; Mutter et al., 1998b). RGLs are in the polysaccharide lyase families 4 and 11 (Lombard et al., 2014; CAZy, 2015).

RGA (EC 3.1.1.86) catalyzes the hydrolytic cleavage of acetyl groups from rhamnogalacturonan in RGI (Searle-van Leeuwen et al., 1992). RGA belongs to carbohydrate esterase family 12 (Lombard et al., 2014; CAZy, 2015). XGH hydrolyzes the α-1,4-D linkages of xylose-substituted galacturonan moieties in XGA, thereby producing xylose galacturonate dimers (van der Vlugt-Bergmans et al., 2000; Zandleven et al., 2005). XGAs belong to glycosyl hydrolase family 28 (Lombard et al., 2014; CAZy, 2015). Accessory enzymes acting on the lateral chains of RG-I and RG-II include the following: endogalactanases (e.g., EC 3.2.1.89); exogalactanases, AGLs (e.g., EC 3.2.1.22); LACs (e.g., EC 3.2.1.23); AFs (e.g., EC 3.2.1.55); AR (EC 3.2.1.99); exoarabinases; and FAE (EC 3.1.1.73) (de Vries and Visser, 2001).

3.2 Natural cellulosomes

As bacteria and fungi are unable to engulf large particles, these organisms must secrete enzymes (e.g., cellulases) that degrade lignocellulose found in plant cell walls. The approach to plant cell wall deconstruction can be quite different in aerobic versus anaerobic microorganisms. In anaerobes, expressed and secreted cellulases and hemicellulases are incorporated into large multienzyme complexes called cellulosomes. The size of such complexes often exceeds 2 MDa (Bégum and Lemaire, 1996; Bayer et al., 1998, 2004; Shoham et al., 1999; Gilbert, 2007; Smith and Bayer, 2013). In aerobic microorganisms, expressed cellulases and hemicellulases are secreted in high concentrations as free enzymes. Many contain carbohydrate-binding modules or CBMs (Wilson, 2008).

Cellulosomes are supramolecular assemblies typically bound to the exterior surface of the microorganism (Smith and Bayer, 2013). Cellulosomes may be charged with an array of cellulose-degrading enzymes, which act synergistically to deconstruct lignocellulosic materials. These synergistic interactions of cellulosome enzyme complements act to efficiently degrade even recalcitrant crystalline lignocellulosic substrates (Fierobe et al., 2002, 2005). One of the most well-studied cellulosomes is that from the anaerobic thermophilic bacterium *Clostridium thermocellum*. The cellulosome of *C. thermocellum* was first described in the year 1980s (Bayer et al., 1983; Lamed et al., 1983a, 1983b). It contains not only cellulases but also a large array of hemicellulases (Morag et al., 1990; Kosugi et al., 2002) and pectinases (Tamaru and Doi, 2001). These cellulosomes may also include polysaccharide lyases, CEs, and glycoside hydrolases. In some anaerobic bacteria, both cellulosomes and free cellulases are expressed; however, the function of the free enzymes in cellulose degradation is still unknown (Gilad et al., 2003; Berger et al., 2007). Anaerobic fungi also construct cellulosomes. Fungal cellulosomes have been reported for species of the genera: *Neocallimastix, Piromyces,* and *Orpinomyces* (Wilson and Wood, 1992; Ali et al., 1995; Fanutti et al., 1995; Li et al., 1997; Fillingham et al., 1999; Steenbakkers et al., 2001, 2003; Nagy et al., 2007; Haitjema et al., 2013; Wang et al., 2014b).

3.2.1   Cellulosome structure

The cellulosome is an extracellular protein complex on bacterial and fungal cell surfaces that adhere to plant materials and deconstruct plant cell wall lignocellulose (Lamed et al., 1983b). Cellulosomes are composed of several subunits, each of which exhibits modular architecture. Some components of the cellulosome are structural, whereas others are catalytic. The core structural component is known as scaffoldin, which is the framework to which all other subunits attach (Tokatlidis et al., 1991). The catalytic components (e.g., enzymes) of the cellulosome contain noncatalytic dockerin domains (Hall et al., 1988). These dockerin modules bind to cohesins, which are the part of scaffoldins (Tokatlidis et al., 1991; Schaeffer et al., 2002; Carvalho et al., 2003). The cellulosome may consist of more than one dockerin type and more than one cohesin type. Dockerin–cohesin interactions may be used to link enzymes to a scaffoldin complex or may be used to anchor the scaffoldin complex to the cell wall. In either case, the high-affinity protein–protein interaction between dockerin and cohesin is fundamental to the assembly of a functional cellulosome endowed with a diverse array of enzymes integrated into the complex (Carvalho et al., 2003). Cellulosomes will also often feature a noncatalytic module called CBM within the scaffoldin framework. CBM anchors the entire complex onto the plant cell wall (Figure 3.8).

Scaffoldins are large noncatalytic modular cohesin-containing biomolecular complexes that are critical for the assembly of a functional cellulosome and substrate binding via the CBM (Tokatlidis et al., 1991; Salamitou et al., 1992). Scaffoldins have been identified in

Figure 3.8 Basic schematic representation of the *C. thermocellum* cellulosome. (Adopted from Stahl, S.W. et al., *Proc. Natl. Acad. Sci. USA*, 109, 20431–20436, 2012.)

several cellulosome-producing bacteria, including *Acetivibrio cellulolyticus*, *Clostridium cellulolyticum*, *Clostridium cellulovorans*, and *Clostridium josui* (Shoseyov et al., 1992; Kakiuchi et al., 1998; Pagès et al., 1999; Dassa et al., 2012). Cellulosomal enzymes that contain dockerin domains will attach to scaffoldin at cohesin. Dockerin contains highly conserved duplicated segments of approximately 22 amino acids. Each conserved segment is connected by 8–17 amino acids.

Dockerin is typically found as a single domain at the C-terminus of cellulosomal enzymes (Grépinet and Béguin, 1986; Yagüe et al., 1990). The reaction between dockerin and cohesin requires calcium in *C. thermocellum* (Yaron et al., 1995; Choi and Ljungdahl, 1996) and *C. cellulolyticum* (Pagès et al., 1997). The first 12 residues of each duplicated segment resemble the calcium-binding loop in the EF-hand motif (Pagès et al., 1997).

Calcium is essential for dockerin stability. Ca^{2+} plays a role in dockerin folding (i.e., structure) and function (Choi and Ljungdahl, 1996; Lytle et al., 2000). As Ca^{2+} is an essential factor in cohesin–dockerin interactions, it is important in the general structural stability of the cellulosome (Lytle et al., 1999, 2000).

In the presence of divalent cations (e.g., Ca^{2+}), cohesin–dockerin interactions form high-affinity protein–protein interactions. It has been reported that cohesin–dockerin (Type I) interactions have a dissociation constant on the order of 10^{-10} M (Fierobe et al., 1999) and that cohesin–dockerin (Type II) interactions are on the order of 10^{-9} M (Jindou, 2004). It has been suggested that dockerin exhibits some flexibility in the way in which it binds cohesin. Specifically, two different configurations are proposed for cohesin–dockerin binding. This plasticity in binding motifs, in turn, provides flexibility in cellulosome assembly resulting in a range of states through which the incorporation of enzyme activities may occur in the functional cellulosome. Furthermore, plasticity in dockerin–cohesin recognition would also provide alternative modes of interaction between enzymes within the complement and substrate (Carvalho et al., 2007).

Despite the aforementioned plasticity in cellulosome structure and function, cohesins exhibit a surprisingly high degree of amino acid sequence homology regardless of species across the approximately 150-residue cohesin tandem repeats, which are part of the scaffoldin complex (Fujino et al., 1992; Shoseyov et al., 1992; Kakiuchi et al., 1998; Ding et al., 1999). Interestingly, although the similarity is further reflected in the three-dimensional structure of cohesins across species, the cohesin–dockerin interactions still appear to be species-specific (Pagès et al., 1997). Recognition in cohesin and dockerin interactivity appears to be primarily mediated by hydrophobic interactions between a beta-sheet domain in cohesin and one of the helices found in the dockerin structure (Spinelli et al., 2000; Lytle et al., 2001; Miras et al., 2002; Schaeffer et al., 2002; Carvalho et al., 2003, 2007). In some cellulosome systems, including that of

Clostridia, the scaffoldin framework, consisting of multiple cohesin sites for binding dockerin-containing enzymes, is itself attached to the cell surface by a second type of cohesin molecule that serves as an anchoring protein (Leibovitz and Béguin, 1996). This anchoring of scaffoldin to the cell surface is the primary mechanism by which the cellulosome associates with the extracellular surface of the cell while binding substrate.

The binding of cellulosome-expressing cells to substrate is also accomplished through scaffoldin. Most of the catalytic components (i.e., enzymes) of a cellulosome do not feature CBM domains. It is CBM on scaffoldin that mediates attachment of the cell to polysaccharide substrates.

Although there are reports that CBMs function to disrupt substrate crystalline structure (Knowles et al., 1987; Din et al., 1994; Wang et al., 2008), it is generally accepted that the primary function of CBMs is to target and bring catalytic domains into proximity of substrate. By facilitating the exposure of the plant cell wall to cellulosomal enzymes, CBMs increase the local concentration of enzymes on target polysaccharides, which, in turn, enhances conversion efficiency and substrate breakdown (Black et al., 1996; Bolam et al., 1998; Hervé et al., 2010).

CBMs are generally located at either the N-terminus or C-terminus of the scaffoldin complex. CBMs may be formed by protein segments from as few as 30 to more than 200 amino acids long (Juge et al., 2002; Abe et al., 2004; Lunetta and Pappagianis, 2014; Peng et al., 2014).

Originally, CBM domains were called as cellulose-binding modules (CBD) because the first proteins discovered were those that are primarily targeted and bound to crystalline cellulose (Tilbeurgh et al., 1986; Gilkes et al., 1988; Tomme et al., 1988a). However, this term was later replaced with the term cellulose-binding module (CBM) to reflect the diversity of structures and ligand specificities exhibited (Boraston et al., 1999).

CBMs are grouped into 67 families within the CAZy database based on amino acid sequence (Lombard et al., 2014; CAZy, 2015). However, others have organized CBMs into three classifications based on binding specificity (Boraston et al., 2004). Type A CBMs interact with flat surfaces of insoluble polysaccharides, including crystalline cellulose. Type B CBMs bind to internal regions of single polysaccharides (e.g., glycan chains). Type C CBMs recognize small saccharides such as monosaccharides, disaccharides, or trisaccharides. It appears that the orientation and position of aromatic residues in the binding sites of CBMs are the primary drivers of specificity (and affinity) for substrate (Simpson et al., 2000). However, other factors including direct hydrogen bonding (Notenboom et al., 2001; Xie et al., 2001; Pell et al., 2003) and calcium-mediated coordination (Bolam et al., 2004; Jamal-Talabani et al., 2004) have also been shown to play an important role in CBM ligand recognition. In some cases, both substrate specificity and mode of enzymatic action may be defined by cooperative activity between CBMs and cellulosomes. For example, family

3c CBM may influence the processivity of GH9 family *endoprocessive* cellulase (Sakon et al., 1997; Irwin et al., 1998; Li et al., 2007b; Burstein et al., 2009; Oliveira et al., 2009). As another example, manipulation of CBM 22 was shown to change the specificity in a GH10 xylanase such that it displayed primarily β-1,4-β-1,3-glucanase activity (Araki et al., 2004).

3.2.2   Biological functions of cellulosomes

It is generally accepted that, in nature, cellulosomes (such as those found in anaerobic bacteria), are more efficient than *free* enzyme systems (found in aerobic bacteria and fungi) in the deconstruction of plant structural polysaccharides. This notion is supported by data, which demonstrate, for example, that *C. thermocellum* requires much less protein to solubilize an equivalent amount of crystalline cellulose substrate than it takes for *T. reesei* (Johnson et al., 1982). Indeed, *C. thermocellum* consistently exhibits one of the highest rates of hydrolytic activity on substrate (Lynd et al., 2002). Cellulosomes from *C. thermocellum* demonstrate a 50-fold advantage in substrate reduction efficiency on crystalline cellulose over a *Tricoderma*-free cellulolytic system (Demain et al., 2005).

This advantage may be based on the ability of the cellulosome to maintain spatial proximity between catalytic components and substrate, thereby maximizing enzyme–substrate interactions and potential for synergistic activity of different cellulosomal enzymes on recalcitrant substrates (Fierobe et al., 2002, 2005).

Recruitment of different lignocellulolytic enzymes to a cellulosome either in the presence of different substrates or during the course of deconstruction of a given lignocellulosic material appears to be a critical determinant of hydrolytic efficiency. Both the composition of the enzyme complement and the order or organization of cellulosome-bound enzymes seem to prevent nonproductive binding of breakdown products. This may also optimize the spacing and thus the action of enzymes within the complement to facilitate cooperation rather than competition between enzymes with overlapping specificities. Selected enzymes may feature high-affinity binding and bind to a single site on the substrate, whereas others may interact with multiple substrate-binding sites with differential affinity across sites. For the former case, competition for a limited number of binding sites on a given substrate would be minimized. In the case of the latter, the ability of an enzyme to use different binding sites would permit hydrolysis to continue within a lignocellulosic substrate with heterogeneous structural features. Not only the diversity but the number of different enzyme types within the complement may also be important because the concentration of the primary product may exceed the concentration of the original substrate as decomposition/conversion proceeds. In general, the cellulosome provides flexibility and organization to the overall operation of lignocellulose deconstruction, which, in turn, facilitates synergistic interaction between enzymes. Such synergy is in part based on optimizing the ratio between different enzyme types within the cellulosome and the spatial configuration of the enzyme complement.

3.2.3   *Cellulosome of* Clostridium thermocellum

C. thermocellum is a thermophilic, spore-forming anaerobic bacterium capable of hydrolyzing a range of substrates found within the lignocellulosic biomass. The cellulosome of *C. thermocellum* was the first cellulosome discovered in a microorganism and is one of the most well-studied cellulosome (Bayer et al., 1983, 2008; Lamed et al., 1983a, 1983b; Gilbert, 2007; Fontes and Gilbert, 2010; Kothari et al., 2011; Akinosho et al., 2014). The cellulosome of *C. thermocellum* has been shown to be particularly effective in degrading crystalline cellulose (Lynd et al., 2002). A single extracellular cellulosomal enzyme complex can be greater than 2 MDa in size (Coughlan et al., 1985). In some strains, however, multiple cellulosomes aggregate to form much larger supercomplexes called polycellulosomes. Polycellulosomes can feature a total molecular mass as large as 100 MDa (Mayer et al., 1987). The complex composition of most cellulosomes will vary with carbon source (Bhat et al., 1993). Although *C. thermocellum* degrades cellulose, it also has the potential to degrade a number of other polysaccharides (Spinnler et al., 1986; Zverlov et al., 2002a, 2005a).

Several cellulosome-producing microorganisms can express more than one type of scaffoldin. The primary scaffoldin of *C. thermocellum* is CipA (Lamed et al., 1983a; Gerngross et al., 1993; Kruus et al., 1995). CipA consists of nine cohesins (Type I) that recognize Type I dockerin.

CipA also consists of CBMIIIa, which binds crystalline cellulose and exhibits a broad range of binding specificity for different sites on the cell walls of plants (Tokatlidis et al., 1991; Gerngross et al., 1993; Tormo et al., 1996; Blake et al., 2006; Yaniv et al., 2013). CipA also features a C-terminal Type II dockerin, which is linked by a hydrophilic X-domain and does not recognize CipA cohesin (Type I), which binds to the dockerin domains of cellulosomal enzymes. Instead, it recognizes Type II cohesin found at the N-termini of cell-surface proteins. There is more than one of these cohesin Type II anchoring proteins, including SdbA, Orf2P, and OlpB—which anchor either cellulosomes or free enzymes to the cell (Lemaire et al., 1995; Leibovitz and Béguin, 1996; Leibovitz et al., 1997; Adams et al., 2006; Xu and Smith, 2010) (Figure 3.9).

As stated earlier, cohesin–dockerin interactions are typically species-specific (Pagès et al., 1997); however, there are notable exceptions. For example, Xyn11A dockerin found in *C. thermocellum* will bind several cohesins in *C. josui* with high affinity ($K_d = 10^{-8}$ M) (Jindou et al., 2004). Cohesin I is connected to scaffoldin by an *O*-glycosylated linker segment

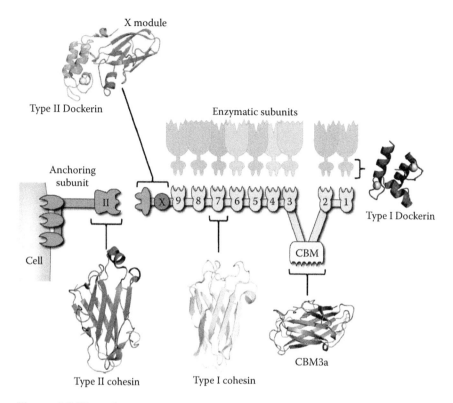

Figure 3.9 The schematic representation of the cellulosome from *Clostridium thermocellum* with the X-ray crystal structures of individual cellulosomal components. (Adopted from Smith, S.P. and Bayer, E.A., *Curr. Opin. Struct. Biol.*, 23, 686–694, 2013.)

of about 20 residues (Gerwig et al., 1989, 1991, 1992, 1993). The linker is composed mostly of proline, serine, and threonine residues. Linkers between cohesins appear to be intrinsically disordered and display a significant degree of conformational flexibility (Hammel et al., 2004, 2005; Bomble et al., 2011; Garcia-Alvarez et al., 2011; Currie et al., 2012, 2013). The protein–protein interaction in the *C. thermocellum* cohesin–dockerin (Type I) complex features a dedicated binding interface with an extensive hydrogen-bonding network and a concomitant set of hydrophobic interactions. These interactions primarily involve only one of the two main helices present in the symmetric dockerin structure and a face of the cohesin module formed by strands 8, 3, 6, and 5 (Carvalho et al., 2003).

Structural and mutagenesis data reveal that dockerin (Type I) contains two, almost identical, cohesin binding interfaces. Residues participating in cohesin recognition at these two interfaces (i.e., a serine–threonine pair at position 11–12 and a lysine–arginine pair at position 18–19) are highly conserved across the majority of *C. thermocellum* dockerins. Conserved sequences in the primary structure at these positions suggest that the two interfaces have similar specificities (Carvalho et al., 2003; Karpol et al., 2008; Garcia-Alvarez et al., 2011). Conversely, plasticity in cohesin–dockerin interactions may reduce steric constraints; so, enzymes may assume different conformations required for substrate degradation (Carvalho et al., 2007). Plasticity may also facilitate dockerin exchanges via recognition of unbound cohesin allowing for a continuous reorganization of the cellulosome.

This provides the cellulosome with a level of structural flexibility that is required to enhance substrate targeting and binding while supporting synergistic cooperation between enzymes within the system (Carvalho et al., 2003, 2007). The lack of interaction between Type I and Type II components of cohesin–dockerin systems supports efficient cellulosome assembly by providing a distinct mechanism for enzyme binding to the structure and for attachment of scaffoldin to the cell surface (Leibovitz and Béguin, 1996).

In terms of binding to lignocellulose substrates, CipA is endowed with CBM3a (CBM family 3). The crystal structure of the *C. thermocellum* CBM3 features a nine-strand β-sandwich fold and one β-sheet presenting a planar topology, which interacts with crystalline cellulose. Within scaffoldin, CBM3 presents as an internal domain consisting of approximately 155 amino acids (Tormo et al., 1996; Yaniv et al., 2013). In comparison to other Type A CBMs, CBM3 series modules bind with high affinity to cellulose (Blake et al., 2006). The attachment of the CBM to cellulose depends on the structural arrangement of the lignocellulosic substrate under attack. CBM3 from *C. thermocellum* has a higher binding affinity (~20-fold higher) for amorphous cellulose than for crystalline cellulose (Morag et al., 1995).

The X module is involved in dockerin stability and cohesin recognition (Adams et al., 2006). The CipA dockerin Type II will only interact with Type II cohesin. A single Type II cohesin may be featured on a cell wall-binding protein, or multiple Type II cohesins may be displayed. Common anchoring proteins SdbA, Orf2P, and OlpB feature 1, 2, and 4 Type II cohesins, respectively (Leibovitz et al., 1997; Adams et al., 2006). Two other CipA anchoring proteins, specifically Cthe_0735 and Cthe_0736—featuring Type II cohesins, have also been described. As many as seven Type II cohesin modules are featured on these anchors, which increase the potential of high-order polycellulosome structures (Raman et al., 2009, 2011).

Although Type II cohesins appear to exclusively function in anchoring cellulosome components to the cell wall, Type I cohesins can be found both within the scaffoldin structure and on the cell surface. In the former case, Type I cohesins serve to bind Type I dockerin domains to associate cellulosomal enzymes to the larger structure. In the latter case, Type I cohesins can act to anchor Type I dockerin-containing enzymes directly to the cell surface.

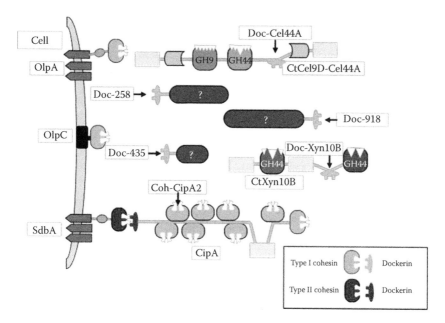

Figure 3.10 Simplified representation of *C. thermocellum cellulosome* with cell-anchored proteins, OlpA, OlpC, and SdbA. The symbol "?" in this figure implies the following: There are three proteins that do not have an assigned function. The dockerin can be located on either the N- or C-terminus of those proteins. (Adopted from Pinheiro, B.A. et al., *Biochem. J.*, 424, 375–384, 2009.)

Two Type I cohesin–containing cell surface anchor proteins (i.e., OlpA and OlpC) have also been identified, which bind individual cellulosome components, such as dockerin Type I enzymes (Salamitou et al., 1994; Pinheiro et al., 2009) (Figure 3.10). Although SdbA, OlpB, and Orf2 anchor proteins in *C. thermocellum* are bound to the peptidoglycan layer of the cell envelope, Type I cohesin-containing OlpA and OlpC bind cell wall polymers in the S-layer of the cell envelope (Lemaire et al., 1995; Zhao et al., 2006a, 2006b; Pinheiro et al., 2009). The evolutionary and physiological implications of these different binding mechanisms are unclear.

With regard to the catalytic core of the cellulosome in *C. thermocellum*, several cellulosomal enzymes have been revealed. Notably, the complex contains numerous EGs (Shinmyo et al., 1979; Ait et al., 1979b; Garcia-Martinez et al., 1980; Ng and Zeikus, 1981; Petre et al., 1981, 1986; Beguin et al., 1983, 1985; Cornet et al., 1983; Schwarz et al., 1986, 1988; Joliff et al., 1986a, 1986b; Soutschek-Bauer and Staudenbauer, 1987; Mel'nik et al., 1989; Hazlewood et al., 1990; Fauth et al., 1991; Jung et al., 1992; Romaniec et al., 1992; Kobayashi et al., 1993; Mosolova et al., 1993; Singh and Akimenko, 1994; Bhat et al., 2001; Zverlov et al., 2003, 2005b).

Other enzymes include the EXs or CBHs. This includes CbhA (formerly Cbh3; component S3) (Tuka et al., 1990; Mel'nik et al., 1991; Singh and Akimenko, 1993; Zverlov et al., 1998); CelS (S8) (Wang et al., 1993; Wang and Wu, 1993); CelK (S5) (Kataeva et al., 1999); and CelO (Zverlov et al., 2002b). The *C. thermocellum* cellulosome employs a cellobiose phosphorylase (Sih and McBee, 1955; Alexander, 1968), a cellodextrin phosphorylase (Sheth and Alexander, 1967, 1969), and two β-glucosidases (bglA and bglB) (Aït et al., 1979a; Gräbnitz et al., 1989, 1991). At least six xylanases are found in this system: XynA, XynB, XynC, XynY, XynZ, and XynD (Grépinet et al., 1988; Morag et al., 1990; Fontes et al., 1995; Hayashi et al., 1997, 1999; Zverlov et al., 2005a). XynY and XynZ are equipped with xylan esterase modules that can remove feruloyl residues from native xylan (Blum et al., 2000). Also present are at least one xyloglucanase (XghA) (Zverlov et al., 2005a), two lichenases (1,3-1,4-β-glucanases) (Schwarz et al., 1985; Schimming et al., 1991), and two laminarinases (1,3-β-glucanases) (Tuka et al., 1990). Evidence also suggests secondary activity of β-xylosidase, β-galactosidase, and β-mannosidase (Kohring et al., 1990). It is important to note that *C. thermocellum* has been shown to break down pectin and thus it likely produces at least one pectin lyase, polygalacturonate hydrolase, pectin methylesterase (Spinnler et al., 1986), chitinase (Chi18A) (Zverlov et al., 2002a), and mannanase (Halstead et al., 1999).

The suite of lignocellulose-degrading enzymes found in *C. thermocellum* illustrates the complex structure and operational flexibility of the cellulosome. The complexity of the catalytic core as well as the multiple options for anchoring to cell wall and substrate binding provide insights into evolutionary pathways through which living systems have sought to convert stored energy bound in raw materials within the environment into usable energy molecules (i.e., simple sugars) for intracellular biochemical reactions that support life.

3.2.4   *Lignocellulolytic system of* Trichoderma reesei

Another evolutionary strategy for accessing energy molecules from more complex carbohydrate precursors does not require the complexity of a large and dynamic cellulosomal structure. Indeed, most aerobic microorganisms, including both fungi (e.g., *Trichoderma reesei*) and bacteria (e.g., *Thermobifida fusca*), produce single enzyme components at high concentrations to deconstruct ligocellulosic materials. Although there is no cellulosome present, many of these enzyme systems do have substrate-binding domains (Wilson, 2008). Substrate-binding domains may be modular and may act synergistically with other enzymes to degrade polysaccharides. Primary and synergistic activities of such enzymes have been most extensively studied in the fungi of the genus *Trichoderma* (Gosh and Gosh, 1992). In particular, the species *Trichoderma reesei* (anamorph of *Hypocrea jecorina*) has served as a model for such systems. As a filamentous meso-philic soft-rot ascomycete fungus, *T. reesei* is widely used in industry as a source for harvesting cellulases and hemicellulases for plant cell wall polysaccharide hydrolysis (Kuhls et al., 1996; Merino and Cherry, 2007a). *T. reesei* is known for its exceptional ability to secrete enzymes and for its ability to grow on a range of substrates (Schaffner and Toledo, 1991). The capacity to produce large quantities of protein and to rapidly switch genes on and off with changing environmental conditions makes *T. reesei* a key player in industrial enzyme production. Certain strains of *T. reesei* can produce >100 grams of extracellular protein per liter of culture (Schuster and Schmoll, 2010). Because of the production efficiency of the industrial strains, *T. reesei* serves as a primary source for cellulase and hemicellu-lase for pulp and paper industries (Buchert et al., 1998), textile industries (Galante et al., 1998), food and feed industries (Hjortkjaer et al., 1986; Roldán et al., 2009), and biofuel production (Kataria and Ghosh, 2011).

Enzyme production in *T. reesei* can be readily controlled and manipu-lated. By altering growth medium (e.g., carbon source) and culture con-ditions (e.g., pH, temperature), enzyme production can be regulated at the transcriptional level both in terms of the types of lignocellulolytic enzymes expressed and the relative quantities of enzymes produced within the bulk culture (Allen and Roche, 1989; Foreman et al., 2003; Juhasz et al., 2005; Stricker et al., 2008; Sipos et al., 2010; Maurya et al., 2012; Coffman et al., 2014). For example, the preferential production of xyl-anases can be induced by increasing lactose concentrations in the culture medium, whereas increased yields of β-xylosidase result after augmenta-tion of the cell medium with xylose (Kristufek et al., 1995; Xiong et al., 2004). Although *T. reesei* has been the industrial standard, studies of its genome have helped scientists identify hydrolytic enzymes in other spe-cies (Martinez et al., 2008).

Bioprospecting has led to the characterization of other species of fungi (and bacteria), which actually express a more diverse array of cel-lulases, hemicellulases, pectinases, and other lignocellulolytic enzymes than *T. reesei* (Martinez et al., 2008).

Indeed, compared to several other fungi (e.g., some white-rot fungi), *T. reesei* has a very small repertoire of genes coding for cellulases, hemicel-lulases, and pectinases (Martinez et al., 2008). As a result, representatives of several of the enzyme families involved in polysaccharide degradation are not found in *T. reesei*, regardless of environmental manipulation of the culture.

Similar to enzyme expression in *T. reesei*, lignocellulolytic enzyme production in other species also depends on the carbon sources and other environmental/culture conditions (Jun et al., 2013). Scientists involved in protein engineering have used these properties of fungal (and bacterial) enzyme expression to facilitate the emergence of modified enzymes with improved properties, such as alkali tolerance, acid tolerance, increased thermostability, and greater catalytic activity (Wang et al., 2005; Nakazawa et al., 2009).

Carbohydrate active enzymes (CAZymes) are enzymes that degrade, modify, or generate glycosidic bonds. These include glycoside hydrolase, which hydrolyzes or rearranges glycosidic bonds; glycosyltransferase (GT), which forms glycosidic bonds; polysaccharide lyase (PL), which nonhydrolytically cleaves glycosidic bonds; and carbohydrate esterase (CE), which hydrolyzes carbohydrate ester bonds. Clusters of genes encoding CAZymes may be found in the genome of *T. reesei*. In general, CAZymes involved in polysaccharide degradation are not distributed randomly in the genome; instead, they are located in clusters. This type of genomic organization seems to be characteristic of many species of Sordariomycetes; interestingly, the number of genes encoding CAZymes in *T. reesei* is below average when compared to the number of CAZyme genes found in other species of the Sordariomycetes (Martinez et al., 2008). For example, the number of *T. reesei* genes encoding glycoside hydrolase is 201, anywhere from 2% to 15% than the number of GH genes found in other species of Sordariomycetes (Martinez et al., 2008; Hakkinen et al., 2012). Furthermore, the *T. reesei* genome has the smallest number of CBM-containing proteins (i.e., 36) within the Sordariomycetes (Martinez et al., 2008). *T. reesei* also contains 22 CEs and 5 PL genes (Hakkinen et al., 2012) as well as 99 GTs (Martinez et al., 2008; Hakkinen et al., 2012)—closer to the average for others within the genus. *T. reesei* lacks several protein families that are critical for cellulose degradation in other species. These enzymes include pectate lyase, pectin esterase, tannase, and feruloyl esterase families. Among the plant-degrading fungi, *T. reesei* has the fewest number of genes for pectinolytic enzymes (e.g., family GH28), and invertase is absent (i.e., family GH32) (Martinez et al., 2008).

Despite limitations of the *T. reesei* genome when compared to other cell wall-degrading fungi, the fungus features a robust cellulolytic system, with genes for several enzyme classes including CBHI/CEL6A and CBHII/CEL7A (Shoemaker et al., 1983; Teeri et al., 1983, 1987); EGs (e.g., EGI/CEL7B, EGII/CEL5A, EGIII/CEL12A, EGIV/CEL61A, and EGV/CEL45A) (Penttilä et al., 1986; Saloheimo et al., 1988, 1994, 1997; Okada et al., 1998); putative EGs (i.e., CEL5B, CEL61B, CEL74A, gene 53731, gene 77284) (Foreman et al., 2003; Martinez et al., 2008; The Regents of the University of California, 2015); eleven β-glucosidases, two of which are well characterized (i.e., BGLI/CEL3A, BGLII/CEL1A) (Barnett et al., 1991; Fowler and Brown, 1992; Takashima et al., 1999; Saloheimo et al., 2002); and 11 βGs, 2 of which are also well characterized (i.e., bgl2/Cel1a and bgl1/ Cel3a) (Barnett et al., 1991; Takashima et al., 1999). There are nine candidate enzymes, which include CEL3B, CEL3D, CEL1B, CEL3C, CEL3E, bgl3i, *T. reesei* gene ID 66832, bgl3j, and bgl3 (Foreman et al., 2003; Ouyang et al., 2006; Martinez et al., 2008).

The aforementioned CBHs (i.e., CBHI/CEL6A and CBHII/CEL7A) have been shown to be processive. CEL6A cleaves cellobiose dimers from the nonreducing end of the cellulose chain, whereas CEL7A also cleaves off cellobiose dimers but from the reducing end (Barr et al., 1996). CEL5B, a GH5 cellulase, features a glycophosphatidylinositol (GPI) anchor at the C-terminus, which serves in enzyme attachment/association with the plasma membrane and fungal cell wall. CEL74A, which was later characterized as a putative XGH (Grishutin et al., 2004) is expressed. In addition, enzymes of the glycoside hydrolase family GH61 have been shown to be part of this system and enhance lignocellulose degradation by oxidative mechanisms (Langston et al., 2011). Several novel candidate cellulolytic enzymes have been identified from the genome of *T. reesei* (Foreman et al., 2003), making it a robust model for studying cellolytic enzyme diversity.

Apart from the cellulases and other cellulose-attacking enzymes, *T. reesei* harbors several genes that encode hemicellulases. This includes six ENs, four of which have been characterized and belong to EN families GH10 (XYNIII), GH11 (XYNI, XYNII), and GH30 (XYNIV) (Tenkanen et al., 1992; Torronen et al., 1992; Xu et al., 1998) as well as three candidate ENs (i.e., XYNV or gene 112392, gene 41248, and gene 69276) (Metz et al., 2011; The Regents of the University of California, 2015). There are also one MAN (i.e., MANI) (Stålbrand et al., 1995), one candidate AXL (gene 69944); one candidate β-1,3-mannanase (gene 71554); six candidate MND enzymes (genes: 5836, 69245, 59689, 57857, 62166, and 71554) (The Regents of the University of California, 2015); one characterized AXE (i.e., AXEI); and three putative AXEs (i.e., AXEII, gene 70021, and gene 54219) (Margolles-Clark et al., 1996d; Foreman et al., 2003; Herpoel-Gimbert et al., 2008; The Regents of the University of California, 2015).

Other hemicellulose-degrading enzymes in *T. reesei* include a candidate cutinase (gene 60489) (The Regents of the University of California, 2015); a XGH (CEL74A) (Grishutin et al., 2004); two AgluAs, one characterized from the family GH67 (GLRI) (Margolles-Clark et al., 1996a) and a candidate AgluA of the GH115 enzyme family (i.e., gene 79606) (Hakkinen et al., 2012); five AFs, including one characterized AF (i.e., ABFI) (Margolles-Clark et al., 1996c) and four candidate AFs (i.e., ABFII, ABFIII, gene 3739, and gene 68064) (Foreman et al., 2003; Herpoel-Gimbert et al., 2008; The Regents of the University of California, 2015); nine AGLs, including three characterized (i.e., AGLI, II, and III) (Zeilinger et al., 1993; Margolles-Clark et al., 1996b) and six candidate AGLs (i.e., gene 27219, gene 27259, gene 59391, gene 75015, gene 55999, and gene 65986) (Metz et al., 2011; The Regents of the University of California, 2015); and two LACs, including one characterized enzyme (i.e., bga1) and one candidate enzyme (gene 76852)

(Seiboth et al., 2005; The Regents of the University of California, 2015). In addition, there are five βXs, including one characterized enzyme (i.e., BXLI) (Margolles-Clark et al., 1996c) and four candidate βX genes (i.e., xyl3b, gene 73102, gene 3739, and gene 68064) (Ouyang et al., 2006; The Regents of the University of California, 2015); two AXEs, one characterized (AESI) (Li et al., 2008b) and one candidate (gene 103825) (Hakkinen et al., 2012); a GE (i.e., CIPII) (Foreman et al., 2003; Li et al., 2007a; Pokkuluri et al., 2011); five candidate β-glucuronidases (i.e., gene 76852, gene 71394, gene 106575, and gene 73005) (The Regents of the University of California, 2015), two α-glucuronidases, the characterized GLRI (Margolles-Clark et al., 1996a) and one candidate gene (gene 79606) (Hakkinen et al., 2012); and lastly five candidate AFUs (gene 69944, gene 72488, gene 5807, gene 111138, and gene 58802) (The Regents of the University of California, 2015). Thus, despite the lack of a formal cellulosome even the most conservative repertoire of genomes among plant cell wall-degrading fungi, features a diverse array of lignocellulolytic enzymes. Expression of extracellularly associated enzymes and secretion of *free* enzymes provide an example of an alternative evolutionary solution to converting complex polysaccharides into smaller energy-rich sugars to drive cellular processes.

3.3   Summary: Natural enzymes

Cellulases are enzymes that mediate cellulolysis. Cellulases are classified into three groups (depending on the reaction catalyzed): carbohydrases, oxidative cellulases, and phosphorylases. Carbohydrases are the dominant and the best studied cellulases. Carbohydrases catalyze the hydrolysis of β-1,4-glucosidic bonds of cellulose or cellooligosaccharides and are grouped into endoglucanases (EG), exoglucanases (EXs), and β-glucosidases (βGs). Oxidoreductive cellulases alone are not cellulolytic, but in concert with other hydrolytic enzymes, they accelerate the rate of cellulose degradation by contributing to the synergetic action mainly by eliminating inhibitory products of the more hydrolytic cellulases. Such enzymes use an oxidative mechanism (Eriksson et al., 1974; Eriksson, 1981) and are found in smaller quantity than primary cellulases. Phosphorylases, such as oxidoreductive cellulases, are not cellulolytic on their own but contribute to accelerated rates of cellulose degradation by cooperating synergistically with other cellulases. Phosphorylases also remove inhibitory products. Phosphorylases depolymerize cellobiose and cellodextrins formed during the reduction of cellulose to fermentable sugars using phosphates. Hemicellulases are either glycoside hydrolases or CEs, which hydrolyze or deacetylate hemicellulose, respectively. The principal enzymes that involve in the breakdown of hemicellulose are xylan-degrading enzymes; however, accessory enzymes also play an important role. Principal enzymes that degrade hemicellulose are hydrolytic backbone-cleaving enzymes that catalyze the hydrolysis of β-1,4 bonds of xylans. These enzymes are classified into three types: endoxylanases (ENs), β-xylosidases (βXs), and exoxylanases (EXYs).

In addition to the main xylan-degrading enzymes, several accessory enzymes are in charge of degrading other less abundant types of xylan backbones (xylan backbone-degrading enzymes) and various substituted xylans (xylan-debranching enzymes). These are the GHs and CEs. Ligninolytic enzymes are oxidative enzymes (i.e., oxidoreductases) that catalyze single-electron oxidation of lignin units, resulting in various nonenzymatic reactions such as bond cleavage. This may be termed as lignolysis. Ligninolytic enzymes are divided into two major families: peroxidases and laccases. Additional accessory enzymes such as oxidases, which enhance the lignolytic activity of the main enzymes are also prevalent. Pectinolytic enzymes or pectinases are an enzyme group that catalyzes the cleavage of pectic substances such as pectin.

The main and most studied pectinases are the polygalacturonate backbone-degrading enzymes. They are grouped into PGs, PMGs, lyases, and PEs (a.k.a., PMEs). There are additional accessory pectinases, which require further characterization. These include rhamnogalacturonan and xylogalacturonan backbone-degrading enzymes and debranching enzymes. They are grouped into PAEs, RGRHs, RGGHs, RGHs, RGLs, RGAs, XGHs, and a host of other accessory enzymes.

Cellulosomes are supramolecular structures that are expressed outwardly on surfaces of microorganisms (Smith and Bayer, 2013). Cellulosomes bind lignocellulolytic subunits (i.e., enzymes) such that synergistic interactions emerge to efficiently deconstruct lignocellulose in plant cell walls (Fierobe et al., 2002, 2005). Naturally occurring cellulosomes are considered as one of the most effective molecular machines for deconstructing plant cell wall biomass. Cellulosomes are composed of both structural and catalytic subunits with a modular architecture. The main structural component is called scaffoldin, which assembles all other subunits and, in many cases, contains an additional substrate-binding component—cellulosome-binding module (CBM) (Tokatlidis et al., 1991; Salamitou et al., 1992). Assembly of a functional cellulosome is mediated by cohesin–dockerin interactions, which can bind dockerin-containing enzymes to cohesin-presenting scaffoldin or can anchor dockerin-containing scaffoldin to cohesin-containing cell wall proteins that are expressed by cellulolytic microorganisms (Hall et al., 1988; Tokatlidis et al., 1991; Schaeffer et al., 2002; Carvalho et al., 2003). Ca^{2+} is essential for cohesin–dockerin interactions and thus is required for the stability of the cellulosomes (Lytle et al., 1999, 2000). The high-affinity protein–protein interactions between dockerin and cohesin allow scaffoldin to be charged with a complement of the enzymes of various classes (Carvalho et al., 2003). Generally, cohesin–dockerin interactions appear

to be species-specific (Pagès et al., 1997). Beyond a suite of lignocellu-
lytic enzymes, cellulosomes will employ the CBMs to ensure that these
enzymes are in proximity with the substrate. This concentrates enzymes
and their catalytic activity onto a substrate to improve the rate of lignocellulosic biomass deconstruction efficiency (Black et al., 1996; Bolam
et al., 1998; Hervé et al., 2010).

The organized sequestration of plant cell wall-degrading enzymes
into a macromolecular assembly and the presence of CBM facilitate cooperative and synergistic interactions against substrate between enzymes
of different classes while ensuring productive adsorption to substrate.
The optimized spacing between the various components of the cellulosome maximizes active site binding across a limited number of binding
sites on the substrate and permits coordinated catalysis by enzymes with
different specificities. This ensures that as new components of the cellulose ultrastructure (e.g., new terminal ends) are exposed by enzymes (e.g.,
endocellulases), another class of enzymes (e.g., exocellulases) is poised to
bind products of the previous reaction.

The best studied (and first characterized) cellulosome is that found in
Clostridium thermocellum, which is known for high-efficiency degradation
of crystalline cellulose (Bayer et al., 1983, 2008; Lamed et al., 1983a, 1983b;
Lynd et al., 2002; Gilbert, 2007; Fontes and Gilbert, 2010; Kothari et al.,
2011). A notable feature in *C. thermocellum* is the aggregation of multiple
cellulosomes into larger supercomplexes, called polycellulosomes (Mayer
et al., 1987). Although there is a significant sequence similarity between
cohesins and dockerins across species, cohesin–dockerin interactions are
generally species-specific. Nevertheless, other cohesin–dockerin interactions are not species-specific (Jindou et al., 2004). Furthermore, there is a
level of plasticity in Type I cohesin–dockerin interactions such that single
cohesin subunit (e.g., cohesin I of CipA) may bind a variety of different
dockerin I-containing enzymes in a given species. The balance between
cohesin–dockerin specificity and plasticity endows the cellulosome with
structural flexibility within the highly organized cellulosome (Carvalho
et al., 2003, 2007). The CBM of *C. thermocellum* is a family 3 Type A CBM,
which robustly binds a broader range of cellulosic substrates than other
type A CBMs (Blake et al., 2006). This CBM also exhibits a higher binding affinity for amorphous cellulose than for crystalline cellulose (Morag
et al., 1995). The *C. thermocellum* cellulosome also features Type I and
Type II cohesin-anchoring proteins (Salamitou et al., 1994; Pinheiro et al.,
2009), both of which bind directly to the exterior bacterial cell surface.
C. thermocellum expresses a large array of cellulases as well as hemicellulases (Morag et al., 1990; Kosugi et al., 2002), pectinases (Tamaru and Doi,
2001), lyases, CEs, and glycoside hydrolases.

Most aerobic microorganisms, including the bacterium *Thermobifida
fusca* and fungi such as *Trichoderma reesei*, (a filamentous fungi), do not

organize their lignocellulolytic enzymes within a cellulosome. Instead, these microorganisms produce lignocellulose-degrading enzymes in high concentrations, which connect to binding modules at the cell surface and act synergistically (Wilson, 2008). Filamentous fungi are a major source for industrial enzymes because of their capacity to secrete high quantities of protein and to grow on a wide range of substrates (Schaffner and Toledo, 1991). *Trichoderma reesei* is an industry standard for its high production of cellulases and hemicellulases (Cherry and Fidantsef, 2003; Merino and Cherry, 2007b). Notwithstanding, *T. reesei* lacks several high-activity cellulases, hemicellulases, and pectinases that are found in other fungi and bacteria (Martinez et al., 2008). Genetic modification of lab strains of *T. reesei* may allow expression of foreign enzymes (Seiboth et al., 2012).

Acknowledgment

The author would like to acknowledge Ms. Inés Cuesta-Ureña, a former graduate student in his lab whose master's thesis (2014) formed the foundation for developing sections of this chapter.

chapter four

Engineered enzymes and enzyme systems

Natural systems for deconstructing lignocellulosic substrates are highly efficient for the roles played by their host microorganisms in vibrant ecosystems. However, natural lignocellulolytic enzymes have limitations when used in industrial processes, and cellulosomes are too complex to fully replicate in any meaningful way within a laboratory or industrial setting. The use of live cultures of fungi or bacteria to degrade pulp has met with some success; however, there are other enzyme-mediated processes that are more efficient when equipped with abiotic enzyme systems. When compared to straight chemical catalysis, *abiotic biocatalysis* has enormous advantages. One of the greatest advantages is the reduced need to handle and dispose hazardous chemicals. Thus, it is not surprising that industrial catalysis has become increasingly dependent on enzymes. Still, the majority of naturally occurring enzymes are not optimized for industrial applications. Some lack the useful thermal profile. Others are unstable with fluctuations in pH and/or salinity. Still others require too many cofactors and very specialized conditions for optimal activity. Multiple parameters must be considered when selecting an enzyme for industrial applications. Indeed, few truly natural enzymes are used in industry today. Most enzymes have been condition adapted by developing lab strains of industrial enzyme-producing fungi and bacteria. Other industrial enzymes have been engineered by direct modification of enzyme's primary structure via genetic manipulation to create enzymes with selected properties (Burton et al., 2002).

4.1   Bioprospecting and metagenomics

The search for superior enzymes has been a robust area of research over the past few decades. To overcome the limitations of naturally occurring enzymes, scientists and engineers have used multiple approaches to develop economically viable biocatalysts with desirable characteristics. Early on, the most practical method was to screen microbial culture for the expression of enzymes with desired activities. Laboratory cultures permit extended and reproducible growth. This allowed phenotypic and genotypic characterization of known elements (Ferrés et al., 2015). However, only a small subset of microorganisms can be readily cultivated in a laboratory setting. Thus, scientists began to place more emphasis on bioprospecting and metagenomic approach. Under this strategy, it was anticipated that the biodiversity found in nature would ultimately gift a set of highly efficient enzymes with specific catalytic properties for any industrial application. By developing large metagenomics libraries based on sequencing of entire genomes from environmental samples followed by DNA extraction, fragmentation/gene amplification, and cloning (Handelsman et al., 1998; Srivastava et al., 2013), a faster approach to discovering the *ideal* enzyme for a particular application was anticipated. Although some success was achieved using this approach, the amount of data generated was, in many cases, overwhelming. Screening of large metagenomic libraries was more challenging than initially expected. Moreover, the activity of an enzyme selected through this process may not be as robust once produced in an artificial protein expression system and the sole protein may be optimally active only when working synergistically with other enzymes or cellular or environmental components.

4.2   Enzyme engineering

A more recent and promising alternative to the development of *ideal* industrial enzymes is protein engineering. Using methods in molecular biology, protein biochemistry, and computational biology, scientists have sought to understand the molecular underpinnings of specific enzyme functionalities to generate novel enzymes featuring the most desirable traits from one or more natural or previously modified model enzymes. Two general approaches in protein engineering have been employed over recent decades with varying levels of success: *rational design* and *directed evolution*. (A third approach for protein improvement based on statistical analysis is also used; however, it is not as prevalent and will not be emphasized here). In rational design, knowledge of enzyme structure (primary, secondary, tertiary, and quaternary) is required, and the catalytic mechanism must be known (Johnsson et al., 1993; Pleiss, 2012). Alternatively, in directed evolution, conditions are altered, and artificial selection is employed to direct changes in emerging populations of enzyme-expressing microorganisms. This approach relies on screening samples from cultures after random mutagenesis, molecular recombination, or focused mutagenesis (Packer and Liu, 2015) (Figure 4.1; *left* and *middle* panels).

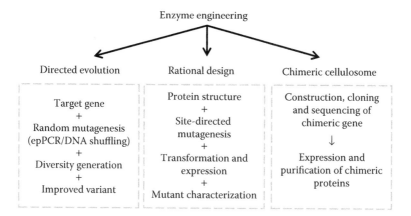

Figure 4.1 Routes to advancement in cellulase enzyme technology. (Adopted and modified from Mohanram, S. et al., *Sustain. Chem. Process.*, 1, 15, 2013.)

4.2.1 Rational design

With the development of polymerase chain reaction (PCR) for amplifying genomic DNA and with the development of methods in generating recombinant DNA, site-directed/specific mutagenesis became possible. Concurrently, methods in biomolecular imaging and modeling were rapidly under development. Using biochemical data, protein structures' (e.g., from X-ray crystallography and protein NMR) biomolecular modeling, and molecular dynamics simulations, it became possible to make reasonable predictions about how changes in protein primary structure would result in changes in three-dimensional conformation (i.e., tertiary structure) and, ultimately, in function. Modifications to genes forcing substitutions, insertions, or deletions in the amino acid sequence at specific positions within protein primary structure ushered in the era of rational design. It is clear that success in rational design is dependent on reliable information about the enzyme structure, function, and mechanism of action. The process of rational design involves (1) choosing a suitable enzyme about which adequate information regarding structure, function, and mechanism is available; (2) identifying amino acid sites that when changed will likely result in structural alteration that produce the desired changes in function; and (3) characterizing the expressed mutants via purification, sequencing, and enzyme activity assays after each round of mutagenesis (Johnsson et al., 1993; Pleiss, 2012). With adequate information regarding structure, function, and mechanism about a target enzyme, rational design is probably the easiest and quickest approach to engineering enzymes. Computational modeling and *in silico* experiments are becoming more sophisticated each year making it even easier to make valid predictions on how function will change when an amino acid or group of amino acids is altered in the primary structure of the protein (Tiwari et al., 2012). Validated predictions are more probable when rational design of an enzyme is based on the knowledge of enzyme structure, function, and mechanism from several related species.

An example of rational design application is the enhancement of enzyme thermostability without reducing catalytic activity. Taking into account protein-surface properties and protein-core characteristics, such as core packing and cavity filling (Joo et al., 2010, 2011), it is possible to strategically add cysteine residues for disulfide bonds (or some other modification) that preserve overall three-dimensional structure and function while improving thermotolerance. Other computational methods, including the development of designer algorithms, the ability to calculate free energy changes, and the production of molecular dynamic simulations not only serve to predict what structure–function changes may result from a directed-mutagenesis but also to predict what mutations may readily occur in genes (and thus proteins) in natural and laboratory systems that are not subject to direct genetic manipulation (Desjarlais and Clarke, 1998).

Rational design is not limited to enhancing a function (or diminishing a function) in one protein. Rational design can be applied to other biomolecular interactions such as pharmaceutical design, enhancement of molecular docking systems, biosynthesis, and development of nanotechnology. Recent methods such as fragment molecular orbital (FMO) calculations and three-dimensional quantitative structure–activity relationship modeling with comparative molecular field analysis (3D-QSAR CoMFA) are further enhancing the utility of rational design (Zhang et al., 2008). Structural databases for protein primary sequence, secondary structure, and 3D conformational information have been developed and rapidly populated over the past two decades. Linking protein databases (e.g., Protein Data Bank [PDB]) to DNA sequence databases (e.g., GenBank) has significantly improved the ability to understand how evolution and genetic mutation (natural or engineered) lead to changes in individual amino acids within a protein and how those amino acid changes impact structure and function. Internet-based computational tools and analytical software packages are now numerous. Combining the use of these tools with information stored in protein databases (such as HotSpot Wizard and 3DM) allows researchers to combine information from sequence/structural searches with functionality to generate mutability maps for target proteins (Pavelka et al., 2009). The commercial 3DM database integrates protein sequence and structural data from

GenBank and the PDB to create comprehensive alignments of protein superfamilies (Kuipers et al., 2010).

Recent methods in protein nuclear magnetic resonance (NMR) relaxation/dispersion coupled with mutagenesis studies provide a powerful new strategy to characterize the effects of controlling (or manipulating) long-range networks of flexible residues on primary active and allosteric binding sites in enzymes and, more generally, enzymatic function (Doucet, 2011).

Site-directed mutagenesis on cellulase was first reported using the *Trichoderma reesei* exo I gene to determine the roles of specific amino acid residues during catalysis (Chen et al., 1987). Rational design was more recently employed to improve the thermostability profiles of βGs from *T. reesei* and *Penicillium piceum* H16 (Lee et al., 2012b; Zong et al., 2015). Other studies have used computational approaches and site-directed mutagenesis to produce a thermostable fungal cellobiohydrolase I (CBHI I) (Cel7A)[*] with a 10°C increase in optimal active temperature (Komor et al., 2012). Rational design was also employed to shift the optimal pH of an endoglucanase (i.e., PvEGIII) from *Penicillium verruculosum* (Tishkov et al., 2013). In another study, rational design was used to enhance catalytic efficiency of EG/EX Cel9A[†] from *T. fusca*; a 40% increase in enzymatic activity against amorphous and crystalline cellulose was achieved (Escovar-Kousen et al., 2004). In yet another study, the actual ratio of products released upon enzymatic action was altered when a single mutation in the active site cleft of the EG-I from *Acidothermus cellulolyticus* was introduced and when the modified enzyme was used to hydrolyze phosphoric acid swollen cellulose (Rignall et al., 2002). Despite these noted achievements, site-directed mutagenesis has only been successful in a few instances to enhance activity of cellulases on insoluble cellulose substrates. One notable success was the modification of EG Cel5A[‡] from *A. cellulolyticus*, which led to a 20% increase in activity on microcrystalline cellulose by decreasing product inhibition (Baker et al., 2005). Deconstruction of crystalline cellulose has been particularly challenging. The mechanism by which cellulases hydrolyze crystalline substrate is not entirely understood. There are insufficient data regarding the mechanism by which a cellulase binds a cellulose within a microfibril structure so that the substrate may be associated within the active site of the enzyme. More work is also required to elucidate mechanisms by which selected CBMs facilitate hydrolysis (Moser et al., 2008; Wang et al., 2008).

Rational design has also been applied in a similar manner to modify hemicellulases. Specifically, properties such as optimal pH and optimal

---

[*] Cel7a is the family-based nomenclature for CBHI.
[†] Cel9A has properties of both the EGs (endoglucanases aka endocellulases) and EXs (exocellulases).
[‡] Endoglucanase (EG) is the common name for the enzymes, whereas Cel5A is the precise nomenclature.

temperature have been shifted to improve enzyme activity in low or high pH conditions (Pokhrel et al., 2013; Xu et al., 2013a, 2013b) and to improve enzyme thermostability (Fonseca-Maldonado et al., 2013; Satyanarayana, 2013). Other studies have focused on improving the catalytic efficiency of selected hemicellulases (Huang et al., 2014; Cheng et al., 2015). Fewer studies are focused on applying rational design to enhance the activity of lignolytic enzymes or pectinases (Xiao et al., 2008; Fang et al., 2014).

Because the most effective strategies to deconstruct lignocellulose often employ multiple enzymes in an *enzyme cocktail*, which relies on synergism, it is critical to note that increasing the individual activity of a single enzyme, or even the individual activities of multiple enzymes within the cocktail, does not necessarily improve the overall efficiency of substrate breakdown. Indeed, the overall synergistic effect may be unchanged or even reduced (Zhang et al., 2000).

Probably, the greatest challenge in rational design is to understand synergism in multienzyme systems and to employ site-directed mutagenesis in a way that enhances synergistic activity. Elucidating the mechanisms by which cooperative enzyme interaction leads to efficient hydrolysis of both crystalline and amorphous regions of cellulose when most individual enzymes only seem to efficiently degrade amorphous substrate has been an ongoing challenge (Chen et al., 2007). Thus, rational design is a powerful tool for modifying the activity profiles of individual enzymes, but it has not resolved the needs for large-scale processing of lignocellulosic materials in which efficient deconstruction relies on synergism between multiple enzymes and enzyme classes. Detailed data about structure–function relationships in cellulase-crystalline substrate activity and interactions between cellulases and other enzymes (e.g., hemicellulases) are still lacking.

Clearly, rational design is a powerful strategy for improving the activity of individual enzymes. Rational design has also been used to engineer metabolic pathways (Eriksen et al., 2014). However, the approach requires a robust characterization of structure, function, and mechanism of target enzymes. For industrial applications, the pure rational design has several limitations primarily due to an incomplete understanding of protein structure and its contribution to function or due to the limited knowledge of protein dynamics and the mechanisms of enzyme synergism in multienzyme cocktails (Ruscio et al., 2009).

4.2.2   *Directed evolution*

Another strategy for engineering individual enzymes is directed evolution. Directed evolution has become another important tool for improving enzyme properties. Similar to rational design, the application of directed evolution can result in the enhancement of enzyme: thermal profile

(Koksharov and Ugarova, 2011; Steffler et al., 2013; Zhou et al., 2015); enantioselectivity (Reetz et al., 1997; May et al., 2000; Kim et al., 2015b); oxidative stability (Oh et al., 2002); general catalytic activity/efficiency (Stemmer, 1994a; Akbulut et al., 2013; Wang et al., 2015); pH profile (Ness et al., 1999; Wang et al., 2005; Melzer et al., 2015); substrate specificity (Glieder et al., 2002; Gupta and Farinas, 2010; Ng et al., 2015); and tolerance or stability toward organic solvents (Moore and Arnold, 1996; Reetz et al., 2010b; Yamada et al., 2015). Similar to rational design, directed evolution can also be applied to engineer metabolic pathways and even whole organisms (Eriksen et al., 2014; Guenther et al., 2014). Directed evolution can be used to generate novel enzyme function (Raillard et al., 2001; Chen and Zhao, 2005). The greatest advantage of directed evolution compared to rational design is that it is independent of the knowledge of enzyme structure and of the interactions between enzyme and substrate. This permits the engineering of enzymes in which the function is not fully understood (Figure 4.2).

Directed evolution is fundamental to the emerging field of *synthetic biology* in which unique suites of enzymes, novel biochemical pathways, or even whole organisms are engineered (Patnaik et al., 2002; Snoek et al., 2015). The process involves iterative cycles of producing mutants and of finding the mutant with the desired properties via screening or selection methods (Arnold, 1998) (Figure 4.3). Directed evolution can be more

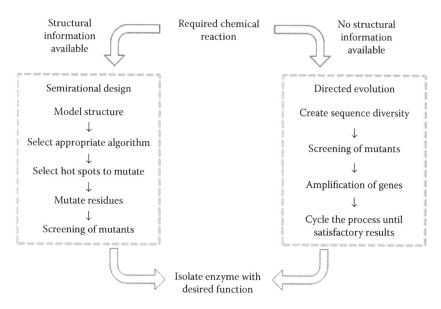

Figure 4.2 Summary of the different processes required by semirational design and directed evolution. (Adopted from Quin, M.B. and Schmidt-Dannert, C., *ACS Catal.*, 1, 1017–1021, 2011.)

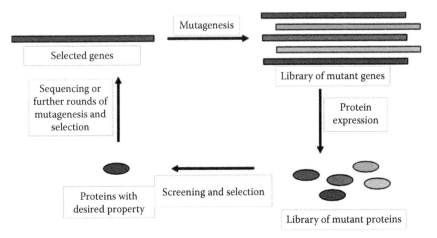

Figure 4.3 General steps of directed enzyme evolution. The gene encoding the protein of interest is mutated to generate a library of mutant genes. Expression of the mutant genes provides the library of mutant proteins. The proteins are screened or selected based on a desired property, and the variants with modified activity are sequenced or used for further rounds of mutagenesis and selection. (Adopted from Tao, H. and Cornish, V.W., *Curr. Opin. Chem. Biol.*, 6, 858–864, 2002.)

efficient than rational design (Gerlt and Babbitt, 2009). This is because directed evolution requires only the knowledge of protein sequence. This is because directed evolution only requires knowledge of the protein primary structure (i.e., the amino acid sequence of an enzyme), which is iteratively altered until the enzyme functions in the desired manner.

Applying directed evolution methods for modifying biomolecules *in vitro* (Mills et al., 1967) and, specifically, for driving nucleic acid (e.g., gene) modifications (Kauffman, 1993) has led to a formal theory for natural molecular evolution and has led to methods for artificial molecular evolution. Early applications of directed molecular evolution were focused on improving enzyme activity, specifically, enhancing the catalytic efficiency of the protease subtilisin E in organic cosolvents (Chen and Arnold, 1993). Since these earliest applications of directed evolution, multiple techniques have been developed and the approach has gained significant traction within the scientific community. One of the most useful outcomes of work in directed evolution has been the development of extensive mutant libraries.

Generally, mutations are introduced into a genome through one of the three primary mechanisms: random mutagenesis, recombination, or focused mutagenesis (Packer and Liu, 2015). Current techniques used to generate mutants by random mutagenesis and recombination include error-prone PCR (Leung et al., 1989) and DNA shuffling (Stemmer, 1994a, 1994b). Error-prone PCR (epPCR) results in the insertion of random point mutations throughout a gene by taking advantage of the low fidelity of the *Taq* polymerase under selected reaction conditions.

These conditions include an increase in magnesium concentrations, manganese supplementation, or the use of mutagenic dNTP analogues (Zaccolo et al., 1996). The DNA shuffling method employs mixing and joining different but related small DNA fragments to generate chimeras. These hybrids or recombinants are essentially new genes that can be tested to determine if viable product can be expressed. This approach has resulted in the development of recombinant or clone libraries based on homologous recombination of the shuffled genes (Crameri et al., 1998).

Directed evolution can also employ nonhomologous recombination (Sieber et al., 2001). Regardless of whether the approach is to use homologous or nonhomologous recombination, large sets of data, typically in recombinant or clone libraries, will need to be accessed to determine which, if any, recombinants possess the modified or novel enzyme properties targeted. Several methods have been developed including look-through mutagenesis (LTM) and combinatorial beneficial mutagenesis (CBMt). LTM is a method developed for rapid screening of amino acid mutations at selected positions within a protein that introduces favorable properties. CBMt is a method used to identify the best ensemble of individual mutations (Hokanson et al., 2011) that results in favorable properties in the protein.

Clearly, experimental design in directed evolution studies requires multiple considerations. Success depends on the mutagenic approach (point mutations versus broader recombination) and the method used to find the resultant mutant enzyme within an extensive combinatorial library (You and Arnold, 1996). Identifying interesting variants within the large combinatorial libraries generated during directed evolution experiments is accomplished in one of two general ways: screening or selection. Screening approaches require assays through which each (and every) individual mutant within the library is tested to determine types of properties that each possesses. Selection approaches set conditions such that only variants possessing desired properties appear (Packer and Liu, 2015). Selection is preferred to screening due to the higher throughput rates (Olsen et al., 2000; Griffiths et al., 2004; Otten and Quax, 2005; Seelig, 2011).

However, screening has an advantage that the difference between the substrate and product after an enzymatic reaction can be determined directly (or, at least indirectly) by a biochemical assay. The disadvantage of screening is that all individual mutants are tested to determine if the desired enzymatic reaction will occur. Screening can be performed via facilitated screening to overcome some of these drawbacks. In facilitated screening, desired mutants can be separated from inactive or undesirable enzymes based on phenotype such as chromospheres or halos. If facilitated screening is not an option, then random screening is used (Taylor et al., 2001). Screening methods may include the use of microtiter plates with colorimetric or fluorescent indicators or the use of quantitative methods in mass spectrometry, capillary electrophoresis, chromatography, or infrared thermography (Wahler and Reymond, 2001; Yang et al., 2010; Despotovic et al., 2012; Yu et al., 2014; Zeng et al., 2015).

Selection approaches follow the *survival of the fittest* phenomenon whereby only those mutants that are capable of utilizing the given substrates under the given conditions are selected. The primary advantage of selection over screening is that the number of mutants that can be simultaneously screened is greater, and variants devoid of target properties are not observed. Thus, evaluating a large mutant library is quick and generally yields a higher number of *hits*. The most advanced screening protocols allow evaluation of up to 10^8 mutants from the library. Selection methods can assess up to 10^{13} mutants with much effort (Packer and Liu, 2015). Selection is based on the fact that mutants with the target or desired enzyme function provide a selective advantage to the host cell over other bacteria in the system bearing wild-type enzymes. Resistance to antibiotics or some other cytotoxic agent is a common way to drive selection (Stemmer, 1994a, 1994b; Siau et al., 2015) as is the use of auxotroph complementation (Smiley and Benkovic, 1994; Jürgens et al., 2000; Griffiths et al., 2004; De Groeve et al., 2009).

Although directed evolution is a method that overcomes some of the drawbacks of rational design, there are still some limitations. Analyses across enzymatic families suggest that major changes in enzyme function require significant changes in the peptide backbone (Matsumura, 2000). However, directed evolution is based on inducing point mutations in which there is a natural bias for transitions over transversions (even in the case of gene shuffling or recombination); thus, the spectrum of substitutions that will emerge in any recombinant library will be limited. Another limitation of directed evolution is the management of a very large number of variants. Even with the aforementioned bias, a library of 10^8–10^{13} recombinants, many which are inactive, requires a significant amount of time and effort to process.

Smaller clone libraries can be generated. However, there is a risk of reducing the diversity that endows directed evolution with its technological power. Methods that yield high-quality libraries while maintaining diversity are the most efficient. To reduce the number of mutants in a library (by at least screening out many of the inactive clones), modifications to substrate and the resulting substrate specificity of variants may be monitored with high-output methods, such as fluorescence-activated cell sorting (FACS). FACS can screen tens of millions of variants in a short amount of time (Bernath et al., 2004; Becker et al., 2008; Fernandez-Alvaro et al., 2011). Directed evolution can also be applied along with the rational design to produce *smarter libraries* (Ba et al., 2013; Teze et al., 2015; Zhang et al., 2015). This *semirational* approach will utilize prior knowledge of primary sequence and/or three-dimensional crystal structure of an enzyme for the development of a more targeted set of mutants, thus reducing the size of the clone library while generating a higher proportion of mutants, which exhibits desirable traits. The semirational approach builds on the strengths of both rational and random variant design to produce smaller but more focused libraries that facilitate efficiency in directed evolution (Reetz et al., 2010a).

Another strategy to narrow the field of potentially useful mutants is to focus on the active site. By limiting the location of changes in the active site to reduce the emergence of inactive mutants and by targeting the active site as a focus for mutant generation, smarter libraries are produced. Primary sequence comparisons can be used to inform which sites and mutations within these regions have a higher likelihood to produce desirable characteristics (i.e., enzyme activities) (Morley and Kazlauskas, 2005; Jochens and Bornscheuer, 2010; Liebgott et al., 2010). Excluding multiple-site mutants also narrows the field because multiple mutations often destabilize proteins (Guo et al., 2004; Drummond et al., 2005; Tokuriki and Tawfik, 2009; Worth et al., 2011). Starting with a very stable protein permits a greater number and range of changes without the loss of function (Bloom et al., 2007; Gupta and Tawfik, 2008).

Other approaches involve the assumption that beneficial mutations are additive (Wells, 1990) and that synergistic effects arising from multiple mutations are rare. Over the past two decades, approaches that allow for an increase in the number of beneficial mutations while minimizing the size of the clone libraries have significantly improved. For example, a single mutation, and certainly no more than four to five substitutions, was typical. However, currently, 30–40 amino acid substitutions are common (Fox et al., 2007; Savile et al., 2010).

4.2.3   Enzyme engineering and biofuels

Enzymes involved in lignocellulosic degradation have been targeted using directed evolution. They are mostly endoglucanases (EGs) and β-glucosidases (βGs) (Arrizubieta and Polaina, 2000; González-Blasco et al., 2000; Kim et al., 2000; Lebbink et al., 2000; Murashima et al., 2002; Catcheside et al., 2003; McCarthy et al., 2004; Wang et al., 2005; Nakazawa et al., 2009; Hardiman et al., 2010; Liu et al., 2010; Liang et al., 2011; Pei et al., 2011; Vu and Kim, 2012; Liu et al., 2013b; Drevland et al., 2014; Lehmann et al., 2014). Directed evolution targeting EGs and βGs is facilitated by the use of soluble or chromogenic artificial substrates, allowing high-throughput screening of variants. Fewer examples of directed evolution targeting improvements in exoglucanases are reported (Wang et al., 2012e; Wu and Arnold, 2013) primarily due to the lack of reliable screening methods.

Directed evolution has achieved limited success in improving activity of individual cellulases. This is primarily due to the difficulties in developing high-throughput screening methods for reactivity on insoluble cellulosic substrates (Zhang et al., 2006b).

Still, directed evolution has been applied to developing cellulases using artificial substrates. Based on these screening protocols, selected cellulases have shown improvement in desirable traits. However, few of these cellulases have shown appreciable improvement on natural substrates (Lin et al., 2009; Nakazawa et al., 2009; Hardiman et al., 2010). With the development of more efficient, high-throughput screening methods on natural substrates, applying directed evolution to the development of high-performance cellulases will become a major thrust in enzyme design (Zhang et al., 2006b; Liu et al., 2010). Such developments are ongoing. Advances in automated microplate spectroscopy now allow high-throughput screening of enzymatic activity on lignocellulosic substrates; these systems will significantly improve cellulase engineering (Chundawat et al., 2008; Navarro et al., 2010; Song et al., 2010; Bharadwaj et al., 2011).

Directed evolution has also been applied to develop high-performance hemicellulases, including those with higher thermostability (Singh et al., 2014; Zheng et al., 2014); different pH optima (Ruller et al., 2014); or overall enzymatic efficiency (Wang et al., 2013; Du et al., 2014).

Directed evolution has also been used for improvement of ligninolytic enzymes and pectinases; however, as with rational design, ligninolytic and pectinolytic enzymes are not well studied (Solbak et al., 2005; Garcia-Ruiz et al., 2012; Liu et al., 2013a; Viña-Gonzalez et al., 2015; Zhou et al., 2015). In general, ligninolytic enzymes and pectinases are understudied when compared to cellulases and hemicellulases. However, the importance of lignin degradation in cellulosic bioethanol production and recent focus on lignin chemistry and lignin separation technologies (Ceballos et al., 2015) may usher in new efforts to develop designer lignin (and pectin)-degrading enzymes.

The high cost of enzymes is a major challenge in producing cost-competitive cellulosic biofuel. It is also a driver for engineering more efficient enzymes for industrial production of bioethanol. In most first-generation (e.g., corn-based) and second-generation (i.e., cellulosic) ethanol production operations, large amounts of enzymes, such as amylases and cellulases, respectively, are required to produce ethanol. In the case of cellulosic ethanol, a multitude of enzyme classes may be required (depending on the feedstock) to break down cellulose into fermentable sugar (Merino and Cherry, 2007a; Klein-Marcuschamer et al., 2012). Thus, it is essential to develop high-performance (and cost-effective) enzymes that can readily degrade lignocellulosic substrate.

Bioprospecting for high-performance natural lignocellulolytic enzymes and enzyme engineering are the ways in which the scientific community has been addressing this need. Common strategies include genome mining in sequenced microbial genomes (Ahmed, 2009; Davidsen et al., 2010); metagenome screening (Handelsman et al., 1998; Srivastava et al., 2013); bioprospecting in extremo- or mesophilic fungi and bacteria (Schiraldi and De Rosa, 2002; Kumar et al., 2011a); and engineering enzymes with properties such as higher efficiency, increased thermostability, and greater tolerance to end-product inhibition.

4.2.4   New technologies for enzyme-mediated lignocellulose deconstruction

Protein engineering is a well-established technology for modifying the properties of enzymes and as new methods and technologies are developed, the path to a more sustainable cellulosic biofuel moves forward. Multiple studies have been published on the advances in protein engineering (Peters et al., 2003; Kazlauskas and Bornscheuer, 2009; Turner, 2009; Bornscheuer et al., 2012; Davids et al., 2013). Strategies for enhancing enzyme-mediated lignocellulose deconstruction include improving the properties of individual cellulases and hemicellulases and synergy engineering through which enzyme cocktails or artificial cellulosomes are designed for maximizing the synergistic effects of multienzyme complements bound to an engineered scaffold/platform (Zhou et al., 2009; Mohanram et al., 2013; Ji et al., 2014; Hu et al., 2015).

To date, there has been some success in engineering cellulases to improve thermostability (Heinzelman et al., 2009a, 2009b, 2010; Komor et al., 2012; Smith et al., 2012; Wu and Arnold, 2013; Trudeau et al., 2014).

The engineering of hemicellulases has been slower. One reason for this is that commercial enzyme cocktails are typically applied to acid-pretreated biomass in which the high-temperature acid washes have already degraded a significant part of the hemicellulose component (Pedersen et al., 2011). Although acid pretreatment (or alkaline pretreatment) of feedstock is effective, this approach requires the handling, use, and disposal of hazardous chemicals. Reducing the need for hazardous chemical management by developing a robust hemicellulase-based component to enzyme cocktails predominantly composed of cellulases may offset this need, providing a *greener* and safer alternative to cellulosic feedstock breakdown. Moreover, a ligninolytic component to an enzyme treatment would open up opportunities for lignin-rich feedstocks, which are currently not amenable for cost-efficient bioethanol production. Careful design of a comprehensive enzyme-mediated feedstock deconstruction method with selected ligninolytic enzymes may act to preserve (rather than precipitate) lignin while still separating out cellulose. Preservation of certain lignin components during feedstock deconstruction would generate valuable precursors that can be converted into lignin coproducts, which in turn could be sold to offset bioethanol production costs, thus lowering the per-gallon production cost of cellulosic bioethanol (Ceballos et al., 2015).

As the global energy consumer base continues to be heavily reliant on fossil fuels for transportation, heating, and other energy needs, it is clear that using the most abundant polymer waste product on Earth—namely, lignocellulosic material—to meet the liquid fuel demands has not been achieved. Still, research marches on in an attempt to resolve the major bottlenecks in bringing cellulosic ethanol to the forefront of the biofuels market. Even though discovering novel high-performance enzymes through bioprospecting and engineering enzymes are the two approaches to solving the *enzyme problem*; another approach is to develop ways to protect these enzymes in industrial settings, to increase catalytic efficiency of enzymes during production, and to maximize the longevity of an enzyme's catalytic lifespan.

To this end, several strategies have been developed including mimicking natural cellulosomes by engineering artificial cellulosomes or *mini-cellulosomes*, which feature components from natural systems (e.g., CBMs) that are modified for use in bulk substrate deconstruction; immobilizing lignocellulolytic enzymes on fixed platforms (either on panels or in columns); and, more recently, employing mobile enzyme sequestration platforms to facilitate feedstock deconstruction (Mitsuzawa et al., 2009; Ceballos et al., 2014).

4.2.4.1 *Artificial cellulosomes*

Hundreds of millions of years of evolutionary history (~700 mya for land plants) have endowed microorganisms with refined machinery for deconstructing cellulosic biomass. Apart from chemical degradation, which necessitates the handling and disposal of hazardous chemicals, scientists have attempted to break down cellulosic materials by mimicking the mechanisms of microorganisms by either developing enzyme cocktails or by engineering artificial cellulosomes.

As complete cellulosomal structures are quite complex (and not even completely understood), some scientists have tried to take selected components of the cellulosomal structure and to construct *mini-cellulosomes*. Attempts to construct artificial cellulosomes or mini-cellulosomes are based on the idea that enzyme activity, particularly multienzyme action, on lignocellulose may be enhanced (compared to free enzyme cocktails) if some of the synergistic effects of cellulosomes can be reconstructed in an engineered system. The simplest constructs include designing enzymes with a CBM domain to enhance catalysis by facilitating substrate binding. More complex systems seek to capture the benefits of a scaffold (e.g., scaffoldin) to bind multiple enzymes, which may act in a synergistic manner as in nature. Efforts include making hybrid scaffoldins using protein

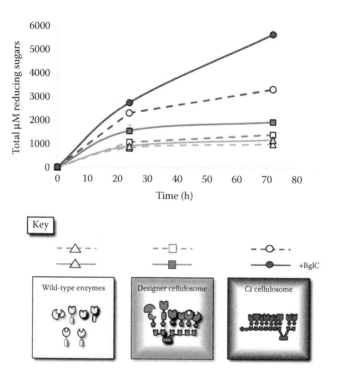

Figure 4.4 Sugar reduction efficiency of designer cellulosome. Sugar reduction efficiency on wheat straw is greatest with a natural cellulosome (blue; right panel). However, an artificial hexavalent cellulosome construct (pink; middle panel) shows greater sugar reduction efficiency than free enzymes in solution (yellow; left panel). Sugar reduction assays were augmented with, a β-glucosidase, BglC (solid lines). Without BglC, sugar reduction efficiency was decreased (dashed lines) for all three systems. (Adopted from Morias, S. et al., *mBio*, 3, 1–11, 2012.)

from more than one bacterial species (Molinier et al., 2011) and genetically modifying yeast to express additional cohesins so that they are capable of binding a larger spectrum of lignocellulolytic enzymes with distinct dockerins (Goyal et al., 2011).

Much of the artificial cellulosome work has been done using the cellulosome of *C. thermocellum* as the model system (Garcia-Alvarez et al., 2011; Molinier et al., 2011; Deng et al., 2015); however, others have focused on other cellulolytic bacteria, including the thermophilic bacterium *Thermobifida fusca* (Mori et al., 2013; Davidi et al., 2016). Many of these artificial cellulosomes exhibit enhanced sugar reduction efficiency over free enzymes (Figure 4.4). However, the magnitude of enhancement does not typically reach the level of efficiency seen in natural cellulosome systems. Moreover, sugar reduction assays provide only an indication of potential enhancement of raw or pretreated lignocellulose biomass deconstruction. Although sugar reduction in lab-scale systems demonstrates improved hydrolysis, upscaling these systems to industrial-scale processing of lignocellulosic biomass while retaining the enhanced effects has proved challenging. Another drawback to designer cellulosomes is cost. When compared to engineering enzymes with desirable properties, designing and producing artificial cellulosomes that may be charged with natural or engineered enzymes are just costly. From an economic standpoint, artificial cellulosome technology is not yet in a position to reduce the per-gallon cost of cellulosic ethanol to a level competitive with fossil-based liquid fuel.

Still, design and development of artificial cellulosomes remain as an ongoing area of intense research. Novel designs and enhancements to prototypes are reported on an annual basis in the scientific literature. As innovative designs continue to emerge in this area of research and as the costs of fossil fuels continue to fluctuate with overall increases over time, artificial cellulosome technology may prove to be an economically viable option for lignocellulose biomass deconstruction in the near future. How engineered cellulosomes will compete against other technologies such as enzyme sequestration platform technologies remains to be seen.

4.2.4.2   Immobilization platforms

Another approach to enhancing enzyme-mediated catalysis is to immobilize enzymes on a platform or in a column. Immobilization technologies permit fine control of catalysis because specificity of immobilized enzymes, enzyme stability, and enhancement of catalytic efficiency by managing enzyme diversity and spatial relationships (e.g., enzyme density) can be designed into the system (Mateo et al., 2007; Hernandez and Fernandez-Lafuente, 2011). Combining site-directed mutagenesis with specific mechanisms that adhere enzymes to a surface is the foundation for immobilized enzyme platforms (Hernandez and Fernandez-Lafuente, 2011). This approach stems from successful application of immobilization platforms for microarrays, enzyme-linked immunosorbent assays, and analyte detection assays.

In its most simple configuration, an enzyme is covalently linked to a solid surface through a metal–ligand interaction (or via some intermediate antigen–antibody interaction). The substrate is then passed over the enzyme-equipped surface, facilitating specific substrate–enzyme interactions.

Other platforms may imbed or otherwise tether an extraneous linker on an enzyme to the surface without disrupting the three-dimensional configuration of the catalytic domains of the enzyme (Brena and Batista-Viera, 2006). Immobilization platforms have proven useful for nanomolar detection of small molecules, nucleic acid, or protein using low-volume flow over a surface (Cohen et al., 2015; Xiaochen et al., 2016). However, when processing large quantities of substrate solution over a two-dimensional surface, limitations of this approach are soon realized. For viscous substrate solutions, or *slurries*, exposing substrate to the immobilized enzymes is increasingly inefficient with increasing volumes of slurry per square centimeter of platform (Krishnan et al., 1999). This problem has been resolved to some extent by moving away from two-dimensional *flat* platforms and instead binding the enzymes within a column.

In column designs, a solid, porous matrix may replace a two-dimensional platform allowing the substrate to pass through the matrix thus increasing the probability that substrate will be exposed to enzyme. Some columns are completely artificial with enzymes bound within a resin in the column (Schifreen et al., 1977; McGhee et al., 1984; Taniai et al., 2001). Others have sought to modify naturally occurring forisomes (Visser et al., 2016), protein populations found within the sieve tubes of the phloem system of Fabaceae (e.g., legumes, pea plants, and bean plants).

For detection-based applications, immobilization platforms are highly efficient. Columns have proven equally useful for separation and purification of macromolecules. High-volume applications have been more challenging. Although some platform (e.g., pressurized column) technologies have been developed for enzyme-mediated cellulosic biomass deconstruction, in general, these have not resulted in *breakthroughs* leading to significant enhancement of enzyme-mediated catalysis and, ultimately, low-cost bioethanol products. With two-dimensional platforms, the thick lignin-, cellulose-, and hemicellulose-containing slurries emerging from mechanical disruption and chemical pretreatment of feedstock do not efficiently direct substrate to specific enzymes, and the slurry does not easily mix across the enzyme-laden surface. Columns can partially overcome the *access* problem; however, flow-through rate is limited. Even in pressurized column systems, fouling and the need to periodically dismantle and re-charge the column increases costs of operation

4.2.4.3   Mobile enzyme platforms

Recent work on mobile enzyme sequestration platforms (mESPs) is the cutting edge of enzyme platform design. mESPs overcome the drawbacks of immobilized platforms by *taking enzyme to the substrate rather than requiring that the substrate comes to the enzyme*. During the breakdown of lignocellulosic biomass, especially that which has been acid pretreated, a thick slurry emerges. Ensuring that polysaccharide substrate within the slurry is exposed to enzyme so that it may be converted to reduced sugars is a drawback with immobilized platforms. However, with mESPs in solution, the enzyme complement acts as another component of the slurry, and more adequate mixing can occur. One of the most innovative approaches in the design of mobile platforms is to build thermotolerance into the platform itself.

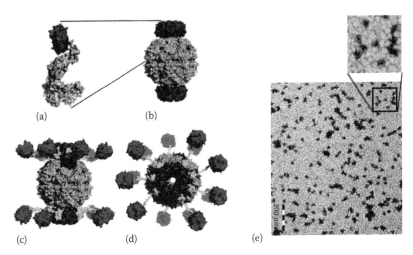

Figure 4.5 Early concept *rosettazyme*. (a) Chaperonin *heat shock* protein sub-unit (grey) is fused to cohesin I from *Clostridium thermocellum* (red) at the apex; (b) double-nonameric ring structure of the functional 18-mer complex serves as the base platform; (c) side view of the platform after it is charged with dockerin-containing cellulolytic enzymes; and (d) top view of the mobile enzyme sequestration platform showing the core. (Adopted from Mitsuzawa, S. et al., *J. Biotechnol.*, 143, 139–144, 2009.); and (e) transmission electron micrograph of a modified mESP. (Adopted from Ceballos, R.M. et al., Improved hydrolysis of pretreated lignocellulosic biomass using mobile enzyme sequestration platforms, in *Recent Advances in Energy, Environment, and Materials*, pp. 47–54, 2014.)

By using a hyperthermophilic chaperonin-based engineered plat-form, enzymes are not only sequestered but also protected from thermal shock. The first prototype of this mESP technology was called the *rosetta-zyme* (Figure 4.5), which showed an enhanced ability over free enzymes to reduce Avicel™, an over-the-counter crystalline cellulose product (Mitsuzawa et al., 2009). A more efficient, second-generation archaeal chaperonin-based mESP system was tested on actual acid-pretreated (and alkaline-pretreated) lignocellulosic feedstock that results in as much as a threefold increase (Figure 4.6) in sugar reduction efficiency over the free enzyme in solution (Ceballos et al., 2014). The key to these designs is that the natural function of chaperonins in the thermal hot spring environment is to protect proteins during pH and temperature fluctuations.

Although the exact mechanism for this protective action is not well understood, the ability to protect cellulases and other enzymes bound to the mESP appears to carry over in the engineered platform system. When acid-pretreated lignocellulosic biomass is treated with a single enzyme or an enzyme cocktail (shown in blue, Figure 4.6), polysaccharides are

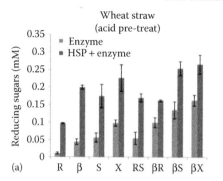
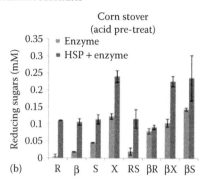

Figure 4.6 Substrate reduction using a mobile enzyme sequestration platform (mESP) on pretreated feedstock biomass. Acid-pretreated feedstock is subjected to a hydrolytic enzyme or enzyme cocktail (*blue bars*) and the same enzyme(s) is bound to a chaperonin-based mobile enzyme sequestration platform (mESP) (*red bars*). Feedstock deconstruction, measured by sugar reduction, in both pretreated wheat straw (a) and corn stover (b) are enhanced by the attachment of enzymes to the mESP. (Adopted from Ceballos, R.M. et al., Improved hydrolysis of pretreated lignocellulosic biomass using mobile enzyme sequestration platforms, in *Recent Advances in Energy, Environment, and Materials*, pp. 47–54, 2014.)

broken down. In general, multiple enzymes in solution result in improved hydrolysis. When the same enzyme or the same complement of enzymes is bound to the protective group II chaperonin-based mobile platform (shown in red, Figure 4.6), a marked enhancement in sugar reduction efficiency over free enzyme in solution emerges.

Mobile enzyme sequestration platforms offer many of the advantages of artificial cellulosomes and platform technology while overcoming major drawbacks associated with immobilizing enzymes on a fixed surface. Furthermore, mESPs can be derived from naturally thermotolerant and acidotolerant protein complexes, increasing their longevity under harsh industrial conditions. mESP design is at the cutting edge of enzyme-mediated lignocellulose deconstruction. The main drawback with mESP technologies is that the efficiency gains by using the mESP must result in a cost saving (e.g., increase in production efficiency or increase in enzyme longevity) that outweighs the cost of adding an additional protein-based component to the process. This is the same issue with 2D platforms and columns. Several emerging technologies may ultimately result in an economically viable enzyme system for deconstructing lignocellulose biomass with such efficiency that the per-gallon cost of producing ethanol is reduced enough to make cellulosic ethanol competitive with fossil-based fuel. These new technologies combined with cost offsets due to revenue generated by the production of sellable coproduct(s) may usher in a new age of alternative liquid fuel production.

4.2.4.4 Future directions

The idea of producing salable coproducts is not new. Indeed, first-generation corn-based ethanol production typically results in the production of *yellow cake* and/or corn oil, both of which have a market value. As separation technologies improve, lignin, the *waste product* of cellulosic ethanol processing, may turn out to be a high-value coproduct. Indeed, a new wave of industrial lignin products are now in demand and can be produced concurrently with bioethanol. Whereas the lignin fraction of lignocellulose-degrading processes was previously considered a hindrance to cellulosic ethanol market viability, industrial lignin precursor production may become the high-value coproduct that makes cellulosic ethanol cost-competitive in the future. Rather than chemically precipitating and burning the lignin fraction, efforts are now underway to separate industrial lignin precursors from lignocellulose while preserving complex carbohydrates (e.g., cellulose) that will ultimately be converted to ethanol (Ceballos et al., 2015).

Nanotechnology is also being applied to enzyme-mediated lignocellulose biomass processing. Designer enzymes that can be enhanced by nanotechnology or that can be recoverable will further drive down cellulosic ethanol production costs. Some efforts are underway to produce an *environmentally friendly* alternative to chemical-based feedstock processing without the use of traditional lignocellulolytic enzymes. In the future, nanocatalysts may replace the need for large multidomain enzymes and enzyme complexes. Such nanocatalysts may be designed to withstand harsh industrial conditions (Candelaria et al., 2017).

Recalling that supplementation or replacement of fossil-based liquid fuel will likely require a combination of alternatives, bringing cellulosic bioethanol to market in conjunction with biodiesel is certainly a laudable aim. Not only is conversion of residual corn oil from corn-based (i.e., first-generation) bioethanol on the horizon, but canola- and algal-based biodiesel are also making gains. Although a detailed treatment of emerging technologies is beyond the scope of this book, efforts to enhance triacylglycerol production in oleaginous microalgae (Gardner et al., 2012) and a rejuvenation of efforts to enhance enzyme-mediated transesterification are setting the stage for bringing biodiesel to the forefront of the liquid fuel market.

4.3 Summary: Enzyme engineering

Bioprospecting microorganisms that may provide unique enzymes capable of degrading cellulose, hemicellulose, lignin, or pectin is ongoing. Anaerobic bacteria, aerobic bacteria, fungi, and even the cyanobacteria and microalgae have contributed to a library of cellulolytic enzymes. Different approaches have been used to discover enzymes with desirable characteristics, including the use of microbial culture collections or metagenomic analysis of unique microbial systems (Satoshi, 1992; Handelsman et al., 1998; Srivastava et al., 2013; Ferrés et al., 2015).

The drawback to these approaches is that novel, useful discoveries are rare, and the overall process is often laborious and slow. Still, the vast biodiversity of the microbial world ensures that additional discoveries, either using traditional culturing approaches or via metagenomics will be heralded in the near future.

However, enzymes found in nature are often optimized for industrial applications so that selected properties are improved (Burton et al., 2002). Engineering enzymes for bioethanol is costly, and the products are one of the major expenditures that producers incur in terms of consumables. The cost of enzymes is considered as one of the main reasons for narrow margins in first-generation bioethanol production and is a major obstacle for the economic viability of cellulosic bioethanol (Merino and Cherry, 2007b; Klein-Marcuschamer et al., 2012). Nonetheless, enzyme engineering, which employs molecular biology techniques and/or computational methods to improve particular enzyme characteristics, may be the strategy that brings cellulosic ethanol to market. Two main approaches are used in engineering cellulolytic enzymes for commercial applications: (1) improving or modifying selected properties of individual enzymes through rational design (Johnsson et al., 1993; Pleiss, 2012) and/or improving enzyme activity by directed evolution (Packer and Liu, 2015) and (2) synergy engineering via optimization of enzyme cocktails or by the construction of multienzyme artificial cellulosomes or enzyme sequestration platforms (Zhou et al., 2009; Mohanram et al., 2013; Ji et al., 2014; Hu et al., 2015).

In rational design, computational models are typically used to predict which amino acid(s) should be inserted, substituted, or omitted by site-directed mutagenesis (Tiwari et al., 2012). Some success has been demonstrated by using rational design to improve cellulase activity (Baker et al., 2005). However, rational design is limited because it requires a comprehensive understanding of the catalytic mechanism of the enzyme being modified—something that is often not available.

Directed evolution overcomes several of the limitations of rational design and has proven to be more successful in the development of designer enzymes (Gerlt and Babbitt, 2009). This is mainly based on the fact that directed evolution requires only knowledge of the protein sequence (i.e., the primary structure) without the need for knowledge of the three-dimensional structure or an understanding of the catalytic mechanism. In directed evolution, repetitive cycles of producing mutants and either screening a mutant library or selecting out undesirable mutants more readily yield enzymes with enhanced or otherwise targeted properties (Arnold, 1998). Either random mutagenesis, random recombination, or focused mutagenesis may be used to produce the mutants (Packer and Liu, 2015). In general, selection methods are preferred to screening methods because of their higher performance (Olsen et al., 2000; Griffiths et al., 2004; Otten and Quax, 2005; Seelig, 2011). Although directed evolution is a powerful technology, one of its main limitations is that only point mutations can be generated. Point mutations will not generally result in significant changes in enzyme function. As significant changes in the polypeptide backbone may be required to achieve the desired modification in enzyme activity, a different approach must be used (Matsumura, 2000).

One strategy to reduce the number of mutants and narrow in those with desired properties is to use directed evolution in conjunction with rational design to produce *smart libraries*; this is known as the *semirational* approach (Ba et al., 2013; Teze et al., 2015; Zhang et al., 2015). Some enzymes that have been successfully modified via directed evolution are the EGs and βGs (Arrizubieta and Polaina, 2000; González-Blasco et al., 2000; Kim et al., 2000; Lebbink et al., 2000; Murashima et al., 2002; Catcheside et al., 2003; McCarthy et al., 2004; Wang et al., 2005; Nakazawa et al., 2009; Hardiman et al., 2010; Liu et al., 2010; Liang et al., 2011; Pei et al., 2011; Vu and Kim, 2012; Liu et al., 2013b; Drevland et al., 2014; Lehmann et al., 2014). Directed evolution has resulted in only moderate success with regard to modifying cellulases. This is primarily due to the difficulties in developing high-throughput screening methods for catalytic activities on insoluble cellulosic substrates (Zhang et al., 2006b). Indeed, enzymes exhibiting enhanced performance on artificial substrates have not generally shown enhanced hydrolysis on natural substrates (Chakiath et al., 2009; Lin et al., 2009; Nakazawa et al., 2009; Hardiman et al., 2010).

Synergy engineering provides an alternative approach to enzyme system design, and it is particularly a useful strategy in the cellulosic ethanol industry. Instead of focusing on individual enzymes, synergy engineering seeks to enhance the cooperative interaction among enzymes when they are present in an enzyme mixture. The principal goal in synergy engineering is to find a combination of enzymes that when used in concert will catalytically outperform any of the individual enzymes in the mixture (in terms of substrate deconstruction). The development of enzyme cocktails, which are simply a selection of enzymes applied free in solution, is the most basic form of synergy engineering. However, synergy engineering may also be applied in the design of artificial cellulosomes or enzyme platforms (i.e., immobilization or mobile platforms). Selecting enzymes for attachment to an artificial cellulosome or designing enzyme complements for binding immobilization platforms or mobile enzyme sequestration platforms that maximize synergistic action on substrate is a preferred strategy when attempting to minimize enzyme loading (Gao et al., 2011; Hu et al., 2011).

Some of the latest research employs several of the aforementioned methods to enhance the deconstruction of lignocellulosic biomass. These techniques may also prove useful for enzymatic transesterification in biodiesel production. As advances in tools and approaches move forward, development of enzyme systems with high-efficiency and high-throughput capabilities in the breakdown of lignocellulosic biomass is expected, leading to a low-cost cellulosic ethanol product. Offsetting production costs by coproduct generation, such as industrial lignin precursors, may ultimately lead to a bioethanol product that can compete with fossil-based liquid fuels.

chapter five

Conclusion

Both biodiesel and bioethanol are promising alternatives to fossil-based liquid fuels. With respect to the latter, first-generation bioethanol (corn-based and sugarcane-based ethanol) results in a *fuel-versus-food* problem. However, second-generation bioethanol (a.k.a., cellulosic ethanol) overcomes this problem by using one of the most abundant waste products on the planet—lignocellulosic biomass. Lignocellulosic biomass can come from bagasse, corn stover, bamboo, trees, wheat straw, rice straw, or a host of other sources. Accessing complex carbohydrates from this biomass (e.g., cellulose) and reducing them to simple, fermentable sugars (e.g. glucose) would provide a significant opportunity to supplement or replace both first-generation bioethanol and fossil-based liquid fuel. However, technologies to effectively deconstruct lignocellulose on an industrial scale in a cost-competitive manner are not yet fully developed.

The depolymerization of carbohydrate substrates in lignocellulosic biomass toward generating fermentable sugars can be done in two different ways: acid hydrolysis and enzymatic hydrolysis. Chemical hydrolysis using highly alkaline solutions and/or high-temperature steam is also used. Acid hydrolysis is mainly performed using sulfuric acid. This requires handling and disposal of hazardous waste and acid has negative effects on downstream processes in the production cycle. Therefore, enzymatic hydrolysis using lignocellulolytic enzymes derived from microorganisms (i.e., bacteria and fungi) is considered a more promising, environmentally friendly technology. Although there is a multitude of different enzymes, no single enzyme (or enzyme class) can completely and efficiently degrade lignocellulose alone. Therefore, enzymatic degradation requires the use of multiple types of enzymes, each with a specificity for hydrolyzing selected components within the lignocellulosic matrix or otherwise for assisting in the deconstruction of lignocellulosic biomass. In both natural and engineering systems, multienzyme systems feature synergistic cooperation in the breakdown of lignocellulosic biomass. In other words, the combined action of multiple enzymes exceeds the additive result of each enzyme working individually on the substrate or even individually in a series of process steps. Synergistic enzyme-mediated lignocellulosic biomass deconstruction may not only act on the catalytic components of the enzyme complement but may also involve features such as carbohydrate-binding modules that act to juxtapose substrate and enzymes (Tomme et al., 1988a, 1998b; Boraston et al., 1999; Gilbert et al., 2013). Other modules or domains within large enzymes mediate cohesin–dockerin interactions, which form the basis for enzyme sequestration in nature either at the microbial cell surface or within large cellulosomes (Tokatlidis et al., 1991; Leibovitz and Béguin, 1996; Schaeffer et al., 2002; Carvalho et al., 2003; Adams et al., 2006; Xu and Smith, 2010). Naturally occurring lignocellulolytic enzymes include cellulases, hemicellulases, lignolytic enzymes, and pectinases. For bioethanol production, cellulases and hemicellulases are the most commonly used enzymes that are used to break down lignocellulose. However, natural enzymes may be modified in whole or in part to optimize catalytic activity for a specific industrial application, such as bioethanol production.

Multiple strategies have been developed to engineer high-performing lignocellulolytic enzymes. Two common approaches in enzyme engineering are rational design and directed evolution. Each of these approaches has its noted advantages but also distinct drawbacks or limitations. Semirational design that employs aspects of each of these strategies is the third approach that attempts to overcome the drawbacks of using either one of these strategies alone. Engineering enzymes to increase catalytic activity, to catalyze reactions in novel ways, or to render the enzyme more resistant to adverse reaction conditions while maintaining catalytic efficiency are not the only focus for engineering enzyme systems. In conjunction with enzyme engineering, or instead of enzyme engineering, a different approach, referred to as synergy engineering, may be used to optimize lignocellulose deconstruction. Synergy engineering focuses more on maximizing the cooperative effects and efficiency of a group of enzymes with distinct activities on substrate. This can be done by simply engineering unique sets of enzyme cocktails, composed of multiple enzymes of different classes, which together work synergistically on a given lignocellulosic substrate. Although enzyme cocktails are the most common approach, more recent research has demonstrated that mixing free enzymes in solution does not maximize cooperation. Indeed, in many naturally occurring cellulolytic systems (i.e., in bacteria and fungi), enzymes are often ordered and bound to structures on the exterior surface of the cell wall. One of the more complex and efficient systems is the cellulosome of *Clostridium thermocellum*. In the cellulosome, sequestration and spatiotemporal ordering of enzymes on a natural framework (i.e., scaffoldin) result in a highly efficient deconstruction of lignocellulosic biomass. As simply using the microorganisms themselves to break down cellulosic feedstock requires infrastructure for culturing anaerobic bacteria, researchers have instead attempted to mimick the cellulosome by engineering artificial cellulosomes. Engineering artificial cellulosomes may include engineering individual enzymes to use in the system and synergy engineering.

The construction of artificial cellulosomes or *mini-cellulosomes* has shown promise in a laboratory setting. However, expression of proteins and assembly of these artificial constructs are not conducive to an industrial setting and do not reduce cellulosic ethanol production costs. Another approach is to develop other scaffold-like frameworks that sequester and bind enzymes. Such platforms can be solid surfaces to which enzymes are bound in an orderly and organized manner. Alternatively, enzymes may be bound in solid, porous columns through which the substrate is passed to facilitate substrate breakdown. Although these enzyme immobilization platforms have proven useful in analyte detection, processing large volumes of viscous slurry from pretreated lignocellulose biomass has proven challenging. This is mainly due to the fact that unlike enzyme cocktails, immobilized enzymes are not free to diffuse throughout the substrate solution. This limits the access to substrate in high flow-through applications. The cutting edge in platform technologies is the mobile enzyme sequestration platforms (mESPs), which exhibit both the advantages of having free enzymes in solution that can readily diffuse through the slurry while retaining the synergistic impacts of sequestering enzymes on a platform. mESP technology may be further enhanced in the future by innovations in nanotechnology, which may increase the stability of these platforms under industrial conditions or even make them recoverable.

Advances in enzyme engineering, including sequestration platform technologies, are at the cusp of providing a highly efficient, high-throughput, and cost-competitive cellulosic ethanol product. With concurrent advances in lignin chemistry and separation technologies, the production of valuable industrial lignin precursors may offset production costs, making cellulosic bioethanol competitive in the world transportation fuel market. Advances in biodiesel production from corn, canola, algae, and other oleaginous feedstocks are moving in sync with advances in cellulosic bioethanol production. Thus, the days of fossil-based liquid fuel that dominated the market may be numbered, and the era of diverse, reasonably priced, and readily accessible alternative biofuels may be on the horizon.

Literature cited

Abe, A., T. Tonozuka, Y. Sakano and S. Kamitori (2004). Complex structures of *Thermoactinomyces vulgaris* R-47 alpha-amylase 1 with malto-oligosaccharides demonstrate the role of domain N acting as a starch-binding domain. *J Mol Biol* **335**(3): 811–822.

Abe, M., Y. Fukaya and H. Ohno (2010). Extraction of polysaccharides from bran with phosphonate or phosphinate-derived ionic liquids under short mixing time and low temperature. *Green Chem* **12**(7): 1274–1280.

Adams, J. J., G. Pal, Z. Jia and S. P. Smith (2006). Mechanism of bacterial cell-surface attachment revealed by the structure of cellulosomal type II cohesin-dockerin complex. *Proc Natl Acad Sci USA* **103**(2): 305–310.

Adav, S. S., C. S. Ng, M. Arulmani and S. K. Sze (2010). Quantitative iTRAQ secretome analysis of cellulolytic *Thermobifida fusca*. *J Proteome Res* **9**(6): 3016–3024.

Ademark, P., R. P. de Vries, P. Hagglund, H. Stalbrand and J. Visser (2001). Cloning and characterization of *Aspergillus niger* genes encoding an alpha-galactosidase and a beta-mannosidase involved in galactomannan degradation. *Eur J Biochem* **268**(10): 2982–2890.

Aden, A., M. Ruth, K. Ibsen, J. Jechura, K. Neeves, J. Sheehan, K. Wallace, L. Montague, A. Slayton and J. Lukas (2002). *Lignocellulosic Biomass to Ethanol Process Design and Economics Utilizing Co-current Dilute Acid Prehydrolysis and Enzymatic Hydrolysis for Corn Stover*. Golden, CO: National Renewable Energy Laboratory.

Adler, E. (1977). Lignin chemistry—past, present and future. *Wood Sci Technol* **11**(3): 169–218.

Agbor, V. B., N. Cicek, R. Sparling, A. Berlin and D. B. Levin (2011). Biomass pre-treatment: Fundamentals toward application. *Biotechnol Adv* **29**(6): 675–685.

Ahmad, M., J. N. Roberts, E. M. Hardiman, R. Singh, L. D. Eltis and T. D. Bugg (2011). Identification of DypB from *Rhodococcus jostii* RHA1 as a lignin per-oxidase. *Biochemistry* **50**: 5096–5107.

Ahmed, N. (2009). A flood of microbial genomes-do we need more? *PLoS One* **4**(6): e5831.

Ahmed, S., A. S. Luis, J. L. Bras, A. Ghosh, S. Gautam, M. N. Gupta, C. M. Fontes and A. Goyal (2013). A novel α-L-arabinofuranosidase of family 43 glycoside hydrolase (Ct43Araf) from *Clostridium thermocellum*. *PLoS One* **8**(9): e73575.

Aït, N., N. Creuzet and J. Cattanéo (1979a). Characterization and purification of thermostable β-glucosidase from *Clostridium thermocellum*. *Biochem Biophys Res Commun* **90**(2): 537–546.

Ait, N., N. Creuzet and P. Forget (1979b). Partial purification of cellulase from *Clostridium thermocellum*. *J Gen Microbiol* **113**(2): 399–402.

Akbulut, N., M. T. Öztürk, T. Pijning, S. İ. Öztürk and F. Gümüşel (2013). Improved activity and thermostability of *Bacillus pumilus* lipase by directed evolution. *J Biotechnol* **164**(1): 123–129.

Akinosho, H., K. Yee, D. Close and A. Ragauskas (2014). The emergence of *Clostridium thermocellum* as a high utility candidate for consolidated bioprocessing applications. *Front Chem* **2**: 66.

Albersheim, P., A. G. Darvill, M. A. O'Neill, H. A. Schols and A. G. J. Voragen (1996). An hypothesis: The same six polysaccharides are components of the primary cell walls of all higher plants. In *Progress in Biotechnology*, J. Visser and A. G. J. Voragen (Eds.), Vol. 14, pp. 47–55. Amsterdam, the Netherlands: Elsevier.

Albersheim, P., H. Neukom and H. Deuel (1960). Über die Bildung von ungesättigten Abbauprodukten durch ein pektinabbauendes Enzym. *Helv Chim Acta* **43**(5): 1422–1426.

Alexander, J. K. (1961). Characteristics of cellobiose phosphorylase. *J Bacteriol* **81**: 903–910.

Alexander, J. K. (1968). Purification and specificity of cellobiose phosphorylase from *Clostridium thermocellum*. *J Biol Chem* **243**(11): 2899–2904.

Ali, B. R. S., L. Zhou, F. M. Graves, R. B. Freedman, G. W. Black, H. J. Gilbert and G. P. Hazlewood (1995). Cellulases and hemicellulases of the anaerobic fungus *Piromyces* constitute a multiprotein cellulose-binding complex and are encoded by multigene families. *FEMS Microbiol Lett* **125**(1): 15–22.

Allen, A. L. and C. D. Roche (1989). Effects of strain and fermentation conditions on production of cellulase by *Trichoderma reesei*. *Biotechnol Bioeng* **33**(5): 650–656.

Alvira, P., E. Tomas-Pejo, M. Ballesteros and M. Negro (2010). Pretreatment technologies for an efficient bioethanol production process based on enzymatic hydrolysis: A review. *Bioresour Technol* **101**(13): 4851–4861.

Amiria, H., K. Karimia and H. Ziloueia (2014). Organosolv pretreatment of rice straw for efficient acetone, butanol, and ethanol production. *Bioresour Technol* **152**: 450–456.

Ander, P. and K.-E. Eriksson (1985). Methanol formation during lignin degradation by *Phanerochaete chrysosporium*. *Appl Microbiol Biotechnol* **21**(1–2): 96–102.

Ander, P., K.-E. Eriksson and H.-S. Yu (1983). Vanillic acid metabolism by *Sporotrichum pulverulentum*: Evidence for demethoxylation before ring-cleavage. *Arch Microbiol* **136**(1): 1–6.

Andreotti, G., A. Giordano, A. Tramice, E. Mollo and A. Trincone (2005). Purification and characterization of a β-D-mannosidase from the marine anaspidean *Aplysia fasciata*. *J Biotechnol* **119**(1): 26–35.

Andrić, P., A. S. Meyer, P. A. Jensen and K. Dam-Johansen (2010). Reactor design for minimizing product inhibition during enzymatic lignocellulose hydrolysis: I. Significance and mechanism of cellobiose and glucose inhibition on cellulolytic enzymes. *Biotechnol Adv* **28**(3): 308–324.

Angel Siles Lopez, J., Q. Li and I. P. Thompson (2010). Biorefinery of waste orange peel. *Crit Rev Biotechnol* **30**(1): 63–69.

Anselmo Filho, P. and O. Badr (2004). Biomass resources for energy in North-Eastern Brazil. *Appl Energy* **77**(1): 51–67.

Antal Jr, M. J. (1996). A traditional solvent pregnant with new application. *Proceedings of the 12th International Conference on the Properties of Water and Steam*, New York, Begell House.

Aracri, E. and T. Vidal (2011). Xylanase- and laccase-aided hexenuronic acids and lignin removal from specialty sisal fibres. *Carbohydr Polym* **83**: 1355–1362.

Arai, M., R. Sakamoto and S. Murao (1989). Different action by two Avicelases from *Aspergillus aculeatus*. *Agric Biol Chem* **53**(5): 1411–1412.

Araki, R., M. K. Ali, M. Sakka, T. Kimura, K. Sakka and K. Ohmiya (2004). Essential role of the family-22 carbohydrate-binding modules for beta-1,3-1,4-glucanase activity of *Clostridium stercorarium* Xyn10B. *FEBS Lett* **561**(1–3): 155–158.

Arantes, V. and J. N. Saddler (2010). Access to cellulose limits the efficiency of enzymatic hydrolysis: The role of amorphogenesis. *Biotechnol Biofuels* **3**: 4.

Ariza, A., J. M. Eklöf, O. Spadiut, W. A. Offen, S. M. Roberts, S. M. Besenmatter, E. P. Friis et al. (2011). Structure and activity of *Paenibacillus* polymyxa xyloglucanase from glycoside hydrolase family 44. *J Biol Chem* **286**(39): 33890–33900.

Armand, S., S. Drouillard, M. Schulein, B. Henrissat and H. Driguez (1997). A bifunctionalized fluorogenic tetrasaccharide as a substrate to study cellulases. *J Biol Chem* **272**(5): 2709–2713.

Arnold, F. H. (1998). Design by directed evolution. *Acc Chem Res* **31**(3): 125–131.

Arora, D. S. and R. K. Sharma (2009). Enhancement in in vitro digestibility of wheat straw obtained from different geographical regions during solid state fermentation by white rot fungi. *Bioresources* **4**(3): 909–920.

Arrizubieta, M. A. J. and J. Polaina (2000). Increased thermal resistance and modification of the catalytic properties of a β-glucosidase by random mutagenesis and in vitro recombination. *J Biol Chem* **275**(37): 28843–28848.

Arvaniti, E., A. B. Bjerre and J. E. Schmidt (2012). Wet oxidation pretreatment of rape straw for ethanol production. *Biomass Bioenergy* **39**: 94–105.

Aspinall, G. O. (1969). Gums and mucilages. *Adv Carbohydr Chem Biochem* **24**: 333–379.

Aspinall, G. O. (1980). Chemistry of cell wall polysaccharides. In *The Biochemistry of Plants (A Comprehensive Treatise)*, J. Preiss (Ed.), Vol. 3, pp. 473–500. New York: Academic Press.

Aspinall, G. O., R. Begbie and J. E. McKay (1962). The hemicelluloses of European larch (*Larix decidua*). Part II. The glucomannan component. *J Chem Soc (Resumed)* 214–219.

Avellar, B. K. and W. G. Glasser (1998). Steam-assisted biomass fractionation. I. Process considerations and economic evaluation. *Biomass Bioenergy* **14**(3): 205–218.

Ayers, A. R., S. B. Ayers and K. E. Eriksson (1978). Cellobiose oxidase, purification and partial characterization of a hemoprotein from *Sporotrichum pulverulentum*. *Eur J Biochem* **90**(1): 171–181.

Ayers, W. A. (1959). Phosphorolysis and synthesis of cellobiose by cell extracts from *Ruminococcus flavefaciens*. *J Biol Chem* **234**(11): 2819–2822.

Azuma, J., F. Tanaka and T. Koshijima (1984). Enhancement of enzymatic susceptibility of lignocellulosic wastes by microwave irradiation. *J Fermen Technol* **62**: 377–384.

Ba, L., P. Li, H. Zhang, Y. Duan and Z. Lin (2013). Semi-rational engineering of cytochrome P450sca-2 in a hybrid system for enhanced catalytic activity: Insights into the important role of electron transfer. *Biotechnol Bioeng* **110**(11): 2815–2825.

Bagby, M. O., G. H. Nelson, E. G. Helman and T. F. Clark (1971). Determination of lignin in non-wood plant fiber sources. *Tappi J* **54**: 1876–1878.

Bai, Y.-Y., L.-P. Xiao, Z.-J. Shi and R.-C. Sun (2013). Structural variation of bamboo lignin before and after ethanol organosolv pretreatment. *Int J Mol Sci* **14**(11): 21394–21413.

Bak, J. S. (2014). Process evaluation of electron beam irradiation based biodegradation relevant to lignocellulose bioconversion. *SpringerPlus* **3**: 487.

Baker, J. O., J. R. McCarley, R. Lovett, C. H. Yu, W. S. Adney, T. R. Rignall, T. B. Vinzant, S. R. Decker, J. Sakon and M. E. Himmel (2005). Catalytically enhanced endocellulase Cel5A from Acidothermus cellulolyticus. *Appl Biochem Biotechnol* **121–124**: 129–148.

Balakshin, M., E. Capanema, H. Gracz, H.-M. Chang and H. Jameel (2011). Quantification of lignin–carbohydrate linkages with high-resolution NMR spectroscopy. *Planta* **233**(6): 1097–1110.

Balan, V., B. Bals, S. P. S. Chundawat, D. Marshall and B. E. Dale (2009). Lignocellulosic biomass pretreatment using AFEX. In *Biofuels: Methods and Protocols*, J. Mielenz (Ed.), pp. 61–77. New York: Humana.

Balan, V., L. da Costa Sousa, S. P. Chundawat, R. Vismeh, A. D. Jones and B. E. Dale (2008). Mushroom spent straw: A potential substrate for an ethanol-based biorefinery. *J Ind Microbiol Biotechnol* **35**(5): 293–301.

Balat, M. (2011). Production of bioethanol from lignocellulosic materials via the biochemical pathway: A review. *Energy Conver Manag* **52**: 858–875.

Balckwelder, D. B., D. J. Muth, E. R. Wilkerson, J. R. Hess and R. D. Perlack (2008). *A Preliminary Assessment of the State of Harvest and Collection Technology for Forest Residues*. Oak Ridge, TN: Oak Ridge National Laboratory.

Baldrian, P. (2006). Fungal laccases—occurrence and properties. *FEMS Microbiol Rev* **30**(2): 215–242.

Ballesteros, M., F. Sáez, I. Ballesteros, P. Manzanares, M. Negro, J. Martínez, R. Castañeda and J. Oliva Dominguez (2010). Ethanol production from the organic fraction obtained after thermal pretreatment of municipal solid waste. *Appl Biochem Biotechnol* **161**(1–8): 423–431.

Banerjee, S., R. Sen, R. A. Pandey, T. Chakrabarti, D. Satpute, B. S. Giri and S. Mudliar (2009). Evaluation of wet air oxidation as a pretreatment strategy for bioethanol production from rice husk and process optimization. *Biomass Bioenergy* **33**(12): 1680–1686.

Barnett, C. C., R. M. Berka and T. Fowler (1991). Cloning and amplification of the gene encoding an extracellular β-glucosidase from *Trichoderma reesei*: Evidence for improved rates of saccharification of cellulosic substrates. *Nat Biotech* **9**(6): 562–567.

Barnett, J. A. (1976). The utilization of sugars by yeasts. *Adv Carbohydr Chem Biochem* **32**: 125–234.

Baron-Epel, O., P. K. Gharyal and M. Schindler (1988). Pectins as mediators of wall porosity in soybean cells. *Planta* **175**: 389–395.

Barr, B., Y. Hsieh, B. Ganem and D. Wilson (1996). Identification of two functionally different classes of exocellulases. *Biochemistry* **35**(2): 586–592.

Barras, D. R., A. E. Moore and B. A. Stone (1969). Enzyme-substrate relations among β-glucan hydrolases. *Adv Chem Ser* **95**: 105–138.

Barreca, A. M., M. Fabbrini, C. Galli, P. Gentili and S. Ljunggren (2003). Laccase/mediated oxidation of a lignin model for improved delignification procedures. *J Mol Catal B: Enzym* **26**(1–2): 105–110.

Bates, F. L., D. French and R. E. Rundle (1943). Amylose and amylopectin content of starches determined by their iodine complex formation. *J Am Chem Soc* **65**: 142–148.

Baxter, L. L., T. R. Miles, T. R. J. Miles, B. M. Jenkins, D. C. Dayton, T. A. Milne, R. W. Bryers and L. L. Oden (1996). *The Behavior of Inorganic Material in Biomass-Fired Power Boilers-Field and Laboratory Experiences: Volume II of Alkali Deposits Found in Biomass Power Plants.* Golden, CO: National Renewable Energy Laboratory.

Bayer, E. A., J.-P. Belaich, Y. Shoham and R. Lamed (2004). The cellulosome: Multienzyme machines for degradation of plant cell wall polysaccharides. *Ann Rev Microbiol* **58**(1): 521–554.

Bayer, E. A., H. Chanzy, R. Lamed and Y. Shoham (1998). Cellulose, cellulases and cellulosomes. *Curr Opin Struct Biol* **8**(5): 548–557.

Bayer, E. A., R. Kenig and R. Lamed (1983). Adherence of *Clostridium thermocellum* to cellulose. *J Bacteriol* **156**(2): 818–827.

Bayer, E. A., R. Lamed, B. A. White and H. J. Flint (2008). From cellulosomes to cellulosomics. *Chem Rec* **8**(6): 364–377.

Bazzi, M. D. (2001). Interaction of camel lens zeta-crystallin with quinones: Portrait of a substrate by fluorescence spectroscopy. *Arch Biochem Biophys* **395**(2): 185–190.

Beale, C. V., D. A. Bint and S. P. Long (1996). Leaf photosynthesis in the C4-grass *Miscanthus* × *giganteus*, growing in the cool temperate climate of southern England. *J Exp Bot* **47**: 267–273.

Beale, C. V. and S. P. Long (1995). Can perennial C4 grasses attain high efficiencies of radiant energy conversion in cool climates? *Plant Cell Environ* **18**(6): 641–650.

Becker, S., H. Hobenreich, A. Vogel, J. Knorr, S. Wilhelm, F. Rosenau, K. E. Jaeger, M. T. Reetz and H. Kolmar (2008). Single-cell high-throughput screening to identify enantioselective hydrolytic enzymes. *Angew Chem Int Ed Engl* **47**(27): 5085–5088.

Beeson, W. T., A. T. Iavarone, C. D. Hausmann, J. H. Cate and M. A. Marletta (2011). Extracellular aldonolactonase from *Myceliophthora thermophila*. *Appl Environ Microbiol* **77**(2): 650–656.

Beeson, W. T., C. M. Phillips, J. H. Cate and M. A. Marletta (2012). Oxidative cleavage of cellulose by fungal copper-dependent polysaccharide monooxygenases. *J Am Chem Soc* **134**(2): 890–892.

Beguin, P., P. Cornet and J. P. Aubert (1985). Sequence of a cellulase gene of the thermophilic bacterium *Clostridium thermocellum*. *J Bacteriol* **162**(1): 102–105.

Beguin, P., P. Cornet and J. Millet (1983). Identification of the endoglucanase encoded by the *celB* gene of *Clostridium thermocellum*. *Biochimie* **65**(8–9): 495–500.

Begum, M. F. and N. Absar (2009). Purification and characterization of intracellular cellulase from *Aspergillus oryzae* ITCC-4857.01. *Mycobiology* **37**(2): 121–127.

Bégum, P. and M. Lemaire (1996). The cellulosome: An exocellular, multiprotein complex specialized in cellulose degradation. *Crit Rev Biochem Mol Biol* **31**(3): 201–236.

Beldman, G., L. A. M. van den Broek, H. A. Schols, M. J. F. Searle-van Leeuwen, K. M. J. van Laere and A. G. J. Voragen (1996). An exogalacturonase from *Aspergillus aculeatus* able to degrade xylogalacturonan. *Biotechnol Lett* **18**(6): 707–712.

Bell, A. and G. F. Wright (1950). The extraction of birch lignins with acetic acid. *J Am Chem Soc* **72**(4): 1495–1499.

Benoit, I., E. G. Danchin, R. J. Bleichrodt and R. P. de Vries (2008). Biotechnological applications and potential of fungal feruloyl esterases based on prevalence, classification and biochemical diversity. *Biotechnol Lett* **30**(3): 387–396.

Bensah, E. C. and M. Mensah (2013). Chemical pretreatment methods for the production of cellulosic ethanol: Technologies and innovations. *Int J Chem Eng* **2013**: 719607.

Bentley, R. and A. Neuberger (1949). The mechanism of the action of Notatin. *Biochem J* **45**: 584–590.

Berger, E., D. Zhang, V. V. Zverlov and W. H. Schwarz (2007). Two noncellulosomal cellulases of *Clostridium thermocellum*, Cel9I and Cel48Y, hydrolyse crystalline cellulose synergistically. *FEMS Microbiol Lett* **268**(2): 194–201.

Berghem, L. E. and L. G. Pettersson (1973). The mechanism of enzymatic cellulose degradation. Purification of a cellulolytic enzyme from *Trichoderma viride* active on highly ordered cellulose. *Eur J Biochem* **37**(1): 21–30.

Berghem, L. E., L. G. Pettersson and U. B. Axiö-Fredriksson (1976). The mechanism of enzymatic cellulose degradation. Purification and some properties of two different 1,4beta-glucan glucanohydrolases from *Trichoderma viride*. *Eur J Biochem* **61**(2): 621–630.

Berlin, A., N. Gilkes, D. Kilburn, R. Bura, A. Markov, A. Skomarovsky, O. Okunev et al. (2005). Evaluation of novel fungal cellulase preparations for ability to hydrolyze softwood substrates—Evidence for the role of accessory enzymes. *Enzyme Microb Technol* **37**(2): 175–184.

Bernath, K., M. Hai, E. Mastrobattista, A. D. Griffiths, S. Magdassi and D. S. Tawfik (2004). In vitro compartmentalization by double emulsions: Sorting and gene enrichment by fluorescence activated cell sorting. *Anal Biochem* **325**(1): 151–157.

Berthelot, M. (1860). Surla fermentation glucosigne du sucre de canne. *Compt Rend Acad Sci* **50**: 980–984.

Berthet, S., N. Demont-Caulet, B. Pollet, P. Bidzinski, L. Cézard, P. Le Bris, N. Borrega, J. Hervé, E. Blondet and S. Balzergue (2011). Disruption of LACCASE4 and 17 results in tissue-specific alterations to lignification of *Arabidopsis thaliana* stems. *Plant Cell Online* **23**(3): 1124–1137.

Bharadwaj, R., A. Wong, B. Knierim, S. Singh, B. M. Holmes, M. Auer, B. A. Simmons, P. D. Adams and A. K. Singh (2011). High-throughput enzymatic hydrolysis of lignocellulosic biomass via in-situ regeneration. *Bioresour Technol* **102**(2): 1329–1337.

Bhat, S., P. W. Goodenough, E. Owen and M. K. Bhat (1993). Cellobiose: A true inducer of cellulosome in different strains of *Clostridium thermocellum*. *FEMS Microbiol Lett* **111**(1): 73–78.

Bhat, S., E. Owen and M. K. Bhat (2001). Isolation and characterisation of a major cellobiohydrolase (S8) and a major endoglucanase (S11) subunit from the cellulosome of *Clostridium thermocellum*. *Anaerobe* **7**(3): 171–179.

Biely, P. (2012). Microbial carbohydrate esterases deacetylating plant polysaccharides. *Biotechnol Adv* **30**(6): 1575–1588.

Biely, P., M. Vrsanska, M. Tenkanen and D. Kluepfel (1997). Endo-beta-1,4-xylanase families: Differences in catalytic properties. *J Biotechnol* **57**(1–3): 151–166.

Binder, A., L. Pelloni and A. Fiechter (1980). Delignification of straw with ozone to enhance biodegradability. *Eur J Appl Microbiol Biotechnol* **11**(1): 1–5.

Bjerre, A. B., A. B. Olesen, T. Fernqvist, A. Ploger and A. S. Schmidt (1996). Pretreatment of wheat straw using combined wet oxidation and alkaline hydrolysis resulting in convertible cellulose and hemicellulose. *Biotechnol Bioeng* **49**(5): 568–577.

Black, G. W., J. E. Rixon, J. H. Clarke, G. P. Hazlewood, M. K. Theodorou, P. Morris and H. J. Gilbert (1996). Evidence that linker sequences and cellulose-binding domains enhance the activity of hemicellulases against complex substrates. *Biochem J* **319**(Pt 2): 515–520.

Blake, A. W., L. McCartney, J. E. Flint, D. N. Bolam, A. B. Boraston, H. J. Gilbert and J. P. Knox (2006). Understanding the biological rationale for the diversity of cellulose-directed carbohydrate-binding modules in prokaryotic enzymes. *J Biol Chem* **281**(39): 29321–29329.

Blanchette, R. A., E. W. Krueger, J. E. Haight, A. Masood and D. E. Akin (1997). Cell wall alterations in loblolly pine wood decayed by the white-rot fungus, *Ceriporiopsis subvermispora. J Biotechnol* **53**(2–3): 203–213.

Bloom, J. D., P. A. Romero, Z. Lu and F. H. Arnold (2007). Neutral genetic drift can alter promiscuous protein functions, potentially aiding functional evolution. *Biol Direct* **2**: 17.

Blum, D. L., I. A. Kataeva, X.-L. Li and L. G. Ljungdahl (2000). Feruloyl esterase activity of the *Clostridium thermocellum* cellulosome can be attributed to previously unknown domains of XynY and XynZ. *J Bacteriol* **182**(5): 1346–1351.

Boerjan, W., J. Ralph and M. Baucher (2003). Lignin biosynthesis. *Annu Rev Plant Biol* **54**: 519–546.

Boisset, C., C. Fraschini, M. Schulein, B. Henrissat and H. Chanzy (2000). Imaging the enzymatic digestion of bacterial cellulose ribbons reveals the endo character of the cellobiohydrolase Cel6A from *Humicola insolens* and its mode of synergy with cellobiohydrolase Cel7A. *Appl Environ Microbiol* **66**(4): 1444–1452.

Bolam, D. N., A. Ciruela, S. McQueen-Mason, P. Simpson, M. P. Williamson, J. E. Rixon, A. Boraston, G. P. Hazlewood and H. J. Gilbert (1998). Pseudomonas cellulose-binding domains mediate their effects by increasing enzyme substrate proximity. *Biochem J* **331**(Pt 3): 775–781.

Bolam, D. N., H. Xie, G. Pell, D. Hogg, G. Galbraith, B. Henrissat and H. J. Gilbert (2004). X4 modules represent a new family of carbohydrate-binding modules that display novel properties. *J Biol Chem* **279**(22): 22953–22963.

Boluda-Aguilar, M. and A. López-Gómez (2013). Production of bioethanol by fermentation of lemon (*Citrus limon* L.) peel wastes pretreated with steam explosion. *Ind Crops Prod* **41**: 188–197.

Bolvig, P. U., M. Pauly, C. Orfila, H. V. Scheller and K. Schnorr (2003). Sequence analysis and characterisation of a novel pectin acetyl esterase from *Bacillus subtilis*. In *Advances in Pectins and Pectinase Research*, A. Voragen, H. Schols and R. Visser (Eds.), pp. 315–330. Dordrecht, the Netherlands: Kluwer Academic Publishers.

Bomble, Y. J., G. T. Beckham, J. F. Matthews, M. R. Nimlos, M. E. Himmel and M. F. Crowley (2011). Modeling the self-assembly of the cellulosome enzyme complex. *J Biol Chem* **286**(7): 5614–5623.

Bonnin, E., K. Clavurier, S. Daniel, S. Kauppinen, J. D. M. Mikkelsen and J.-F. Thibault (2008). Pectin acetylesterases from *Aspergillus* are able to deacetylate homogalacturonan as well as rhamnogalacturonan. *Carbohydr Polym* **74**: 411–418.

Bootten, T. J., P. J. Harris, L. D. Melton and R. H. Newman (2004). Solid-state 13C-NMR spectroscopy shows that the xyloglucans in the primary cell walls of mung bean (*Vigna radiata* L.) occur in different domains: A new model for xyloglucan–cellulose interactions in the cell wall. *J Exp Bot* **55**(397): 571–583.

Boraston, A. B., D. N. Bolam, H. J. Gilbert and G. J. Davies (2004). Carbohydrate-binding modules: Fine-tuning polysaccharide recognition. *Biochem J* **382**(3): 769–781.

Boraston, A. B., B. W. McLean, J. M. Kormos, M. Alam, N. R. Gilkes, C. A. Haynes, P. Tomme, D. G. Kilburn and R. A. J. Warren (1999). Carbohydrate-binding modules: Diversity of structure and function. In *Recent Advances in Carbohydrate Bioengineering*, H. J. Gilbert, G. J. Davies, B. Henrissat and B. Svensson (Eds.), pp. 202–211. Cambridge, UK: Royal Society of Chemistry.

Borneman, W. S., L. G. Ljungdahl, R. D. Hartley and D. E. Akin (1991). Isolation and characterization of *p*-coumaroyl esterase from the anaerobic fungus *Neocallimastix* strain MC-2. *Appl Environ Microbiol* **57**(8): 2337–2344.

Bornscheuer, U. T., G. W. Huisman, R. J. Kazlauskas, S. Lutz, J. C. Moore and K. Robins (2012). Engineering the third wave of biocatalysis. *Nature* **485**(7397): 185–194.

Bose, S., C. A. Barnes and J. W. Petrich (2012). Enhanced stability and activity of cellulase in an ionic liquid and the effect of pretreatment on cellulose hydrolysis. *Biotechnol Bioeng* **109**(2): 434–443.

Bothast, R. J. and M. A. Schlicher (2005). Biotechnological processes for conversion of corn into ethanol. *Appl Microbiol Biotechnol* **67**(1): 19–25.

Bourbonnais, R. and M. G. Paice (1990). Oxidation of non-phenolic substrates. An expanded role for laccase in lignin biodegradation. *FEBS Lett* **267**(1): 99–102.

Bouveng, H. O. (1961). Phenylisocynate derivatives of carbohydrates. *Acta Chem Seand* **15**: 87–95.

Bouveng, H. O. (1965). Polysaccharides in pollen. II. The xylogalacturonan from mountain pine (*Pinus mugo* Turra) pollen. *Acta Chem Scand* **19**: 953–963.

Bowski, L., R. Saini, D. Y. Ryu and W. R. Vieth (1971). Kinetic modeling of the hydrolysis of sucrose by invertase. *Biotechnol Bioeng* **13**(5): 641–656.

Braccini, I., R. P. Grasso and S. Perez (1999). Conformational and configurational features of acidic polysaccharides and their interactions with calcium ions: A molecular modeling investigation. *Carbohydr Res* **317**(1–4): 119–130.

Brena, B. M. and Batista-Viera, F. (2006). Immobilization of enzymes. In *Immobilization of Enzymes and Cells*, J. M. Guisan (Ed.). Totowa, NJ: Humana Press.

Brice, R. E. and I. M. Morrison (1982). The degradation of isolated hemicelluloses and lignin-hemicellulose complexes by cell-free, rumen hemicellulases. *Carbohydr Res* **101**: 93–100.

Brodeur, G., E. Yau, K. Badal, J. Collier, K. B. Ramachandran and S. Ramakrishnan (2011). Chemical and physicochemical pretreatment of lignocellulosic biomass: A review. *Enzyme Res* **2011**: 787532.

Brodie, A. F. and F. Lipmann (1955). Identification of a gluconolactonase. *J Biol Chem* **212**(2): 677–685.

Brosse, N., P. Sannigrahi and A. Ragauskas (2009). Pretreatment of *Miscanthus* × *giganteus* using the ethanol organosolv process for ethanol production. *Ind Eng Chem Res* **48**(18): 8328–8334.

Brown, R. C. (2003). The biorenewable resource base. In *Biorenewable Resources: Engineering New Products from Agriculture*, p. 67. Amess, IA: Iowa State Press.

Brownell, H. H. and J. N. Saddler (1987). Steam pretreatment of lignocellulosic material for enhanced enzymatic hydrolysis. *Biotechnol Bioeng* **29**(2): 228–235.

Bruchmann, E.-E., H. S. Schach and H. Graf (1987). Role and properties of lactonase in a cellulase system. *Biotech Appl Biochem* **9**: 146–159.

Brunow, G. (2006). *Biorefineries—Industrial Processes and Products*. Weinheim, Germany: Wiley-VCH Verlag.

Bubalo, M. C., K. Radošević, I. R. Redovniković, J. Halambek and V. G. Srček (2014). A brief overview of the potential environmental hazards of ionic liquids. *Ecotoxicol Environ Saf* **99**: 1–12.

Buchert, J., J. Oksanen, J. Pere, M. Siika-Aho, A. Suurnakki and L. Viikari (1998). Applications of *Trichoderma reesei* enzymes in the pulp and paper industry. In *Trichoderma and Gliocladium*, C. P. Kubicek and G. E. Harman (Eds.), Vol. 1, pp. 343–364. London: Taylor & Francis Group.

Bugg, T. D., M. Ahmad, E. M. Hardiman and R. Rahmanpour (2011). Pathways for degradation of lignin in bacteria and fungi. *Nat Prod Rep* **28**(12): 1883–1896.

Burstein, T., M. Shulman, S. Jindou, S. Petkun, F. Frolow, Y. Shoham, E. A. Bayer and R. Lamed (2009). Physical association of the catalytic and helper modules of a family-9 glycoside hydrolase is essential for activity. *FEBS Lett* **583**(5): 879–884

Burton, R. A., M. J. Gidley and G. B. Fincher (2010). Heterogeneity in the chemistry, structure and function of plant cell walls. *Nat Chem Biol* **6**(10): 724–732.

Burton, S. G., D. A. Cowan and J. M. Woodley (2002). The search for the ideal biocatalyst. *Nat Biotechnol* **20**(1): 37–45.

Cai, X., X. Zhang and D. Wang (2011). Land availability for biofuel production. *Environ Sci Technol* **45**(1): 334–339.

Camarero, S., S. Sarkar, F. J. Ruiz-Duenas, M. J. Martinez and A. T. Martinez (1999). Description of a versatile peroxidase involved in the natural degradation of lignin that has both manganese peroxidase and lignin peroxidase substrate interaction sites. *J Biol Chem* **274**(15): 10324–10330.

Canam, T., J. R. Town, A. Tsang, T. A. McAllister and T. J. Dumonceaux (2011). Biological pretreatment with a cellobiose dehydrogenase-deficient strain of Trametes versicolor enhances the biofuel potential of canola straw. *Bioresour Technol* **102**: 10020–10027.

Candelaria, S. L., N. M. Bedford, T. J. Woehl, N. S. Rentz, A. R. Showalter, S. Pylypenko, B. A. Bunker et al. (2017). Multi-component Fe-Ni hydroxide nanocatalyst for oxygen evolution and methanol oxidation reactions under alkaline conditions. *ACS Catal* **7**: 365–379.

Canteri-Schemin, M. H., H. C. R. Fertonani, N. Waszczynskyj and G. Wosi-acki (2005). Extraction of pectin from apple pomace. *Braz Arch Biol Technol* **48**: 259–266.

Cao, W., C. Sun, R. Liu, R. Yin and X. Wu (2012). Comparison of the effects of five pretreatment methods on enhancing the enzymatic digestibility and ethanol production from sweet sorghum bagasse. *Bioresour Technol* **111**: 215–221.

Carrard, G., A. Koivula, H. Söderlund and P. Béguin (2000). Cellulose-binding domains promote hydrolysis of different sites on crystalline cellulose. *Proc Natl Acad Sci* **97**(19): 10342–10347.

Carruthers, S. P., C. E. Flint, R. B. Tanter, P. M. Hare and A. R. Staniforth (1991). An assessment of the potential for producing electricity from biomass in the UK (sub-project: Feedstock production). In *Nonfood Uses of Agricultural Products*, House of Lords Select Committee on the European Communities, 7th Report, Session 1990–1991, pp. 80–120.

Carvalho, A. L., F. M. Dias, T. Nagy, J. A. Prates, M. R. Proctor, N. Smith, E. A. Bayer et al. (2007). Evidence for a dual binding mode of dockerin modules to cohesins. *Proc Natl Acad Sci USA* **104**(9): 3089–3094.

Carvalho, A. L., F. M. Dias, J. A. Prates, T. Nagy, H. J. Gilbert, G. J. Davies, L. M. Ferreira, M. J. Romao and C. M. Fontes (2003). Cellulosome assembly revealed by the crystal structure of the cohesin-dockerin complex. *Proc Natl Acad Sci USA* **100**(24): 13809–13814.

Catcheside, D. E. A., J. P. Rasmussen, P. J. Yeadon, F. J. Bowring, E. B. Cambareri, E. Kato, J. Gabe and W. D. Stuart (2003). Diversification of exogenous genes in vivo in *Neurospora. Appl Microbiol Biotechnol* **62**(5–6): 544–549.

CAZy (2015). CAZY database. Retrieved May 20, 2015, from http://www.cazy.org/.

Ceballos, R. M., N. A. Batchenkova, M. K. Y. Chan, A. X. Duffing-Romero, A. E. Nelson, S. Man (2015). Bioethanol: Feedstock alternatives, pretreatments, lignin chemistry, and the potential for green value-added lignin co-products. *J Environ Anal Chem* **2**(5): 1–24.

Ceballos, R. M., R. M. Ceballos Jr., A. Rani, C. T. Morales and N. A. Batchenkova (2014). Improved hydrolysis of pretreated lignocellulosic biomass using mobile enzyme sequestration platforms. In *Recent Advances in Energy, Environment, and Materials*, pp. 47–54., Saint Petersburg, Russia: Saint Petersburg State Polytechnic University, September 23–25, 2014.

Chakiath, C., M. J. Lyons, R. E. Kozak and C. S. Laufer (2009). Thermal stabilization of *Erwinia chrysanthemi* pectin methylesterase a for application in a sugar beet pulp biorefinery. *Appl Environ Microbiol* **75**(23): 7343–7349.

Chandra, R. P., R. Bura, W. E. Mabee, A. Berlin, X. Pan and J. N. Saddler (2007). Substrate pretreatment: The key to effective enzymatic hydrolysis of lignocellulosics? *Adv Biochem Eng Biotechnol* **108**: 67–93.

Chang, M. M., T. Y. C. Chou and G. T. Tsao (1981). Structure, pretreatment and hydrolysis of cellulose. In *Bioenergy*, Vol. 20, pp. 15–42. Berlin, Germany: Springer.

Chang, V. S. and M. T. Holtzapple (2000). Fundamental factors affecting biomass enzymatic reactivity. *Appl Biochem Biotechnol* **84–86**(1–9): 5–37.

Chaubey, M. and V. P. Kapoor (2001). Structure of a galactomannan from the seeds of *Cassia angustifolia* Vahl. *Carbohydr Res* **332**: 439–444.

Chauve, M., H. Mathis, D. Huc, D. Casanave, F. Monot and N. L. Ferreira (2010). Comparative kinetic analysis of two fungal β-glucosidases. *Biotechnol Biofuels* **3**: 3.

Chen, C.-M., M. Ward, L. Wilson, L. Sumner and S. Shoemaker (1987). Toward improved cellulases. Targeted modifications of *Trichoderma reesei* exocellobiohydrolase I/using site-specific mutagenesis. *Abstr Pap Am Chem Soc* **194**.

Chen, K. and F. H. Arnold (1993). Tuning the activity of an enzyme for unusual environments: Sequential random mutagenesis of subtilisin E for catalysis in dimethylformamide. *Proc Natl Acad Sci USA* **90**(12): 5618–5622.

Chen, Y., A. J. Stipanovic, W. T. Winter, D. B. Wilson and Y.-J. Kim (2007). Effect of digestion by pure cellulases on crystallinity and average chain length for bacterial and microcrystalline celluloses. *Cellulose* **14**(4): 283–293.

Chen, Z. and H. Zhao (2005). Rapid creation of a novel protein function by in vitro coevolution. *J Mol Biol* **348**(5): 1273–1282

Cheng, Y.-S., C.-C. Chen, J.-W. Huang, T.-P. Ko, Z. Huang and R.-T. Guo (2015). Improving the catalytic performance of a GH11 xylanase by rational protein engineering. *Appl Microbiol Biotechnol* **99**: 9503–9510.

Cherney, J. H., K. D. Johnson, J. J. Volenec and D. K. Greene (1991). Biomass potential of selected grass and legume crops. *Energy Sour* **13**(3): 283–292.

Cherry, J. R. and A. L. Fidantsef (2003). Directed evolution of industrial enzymes: An update. *Curr Opin Biotechnol* **14**(4): 438–443.

Chiappe, C. and D. Pieraccini (2005). Ionic liquids: Solvent properties and organic reactivity. *J Phys Organ Chem* **18**(4): 275–297.

Chir, J. L., C. F. Wan, C. H. Chou and A. T. Wu (2011). Hydrolysis of cellulose in synergistic mixtures of beta-glucosidase and endo/exocellulase Cel9A from *Thermobifida fusca*. *Biotechnol Lett* **33**(4): 777–782.

Cho, N. S., A. Leonowicz, A. Jarosz-Wilkolazka, G. Ginalska, H. Y. Cho and S. J. Shin (2008). Degradation of a non-phenolic beta-0-4 lignin model dimer by *Cerrena unicolor* laccase and mediators, acetovanillone and acetosyringone. *J Fac Agric Kyushu Univ* **53**: 7–12.

Choi, S. K. and L. G. Ljungdahl (1996). Structural role of calcium for the organization of the cellulosome of *Clostridium thermocellum*. *Biochemistry* **35**(15): 4906–4910.

Choudhary, R., A. L. Umagiliyage, Y. Liang, T. Siddaramu, J. Haddock and G. Markevicius (2012). Microwave pretreatment for enzymatic saccharification of sweet sorghum bagasse. *Biomass Bioenergy* **39**: 218–226.

Chum, H. L., D. K. Johnson, S. Black, J. Baker, K. Grohmann, K. V. Sarkanen, K. Wallace and H. A. Schroeder (1988). Organosolv pretreatment for enzymatic hydrolysis of poplars: I. Enzyme hydrolysis of cellulosic residues. *Biotechnol Bioeng* **31**(7): 643–649.

Chundawat, S. P., V. Balan and B. E. Dale (2008). High-throughput microplate technique for enzymatic hydrolysis of lignocellulosic biomass. *Biotechnol Bioeng* **99**(6): 1281–1294.

Chundawat, S. P., B. Venkatesh and B. E. Dale (2007). Effect of particle size based separation of milled corn stover on AFEX pretreatment and enzymatic digestibility. *Biotechnol Bioeng* **96**(2): 219–231.

Chundawat, S. P. S., G. T. Beckham, M. E. Himmel and B. E. Dale (2011). Deconstruction of lignocellulosic biomass to fuels and chemicals. *Annu Rev Chem Biomol Eng* **2**(1): 121–145.

Clarke, A. E., R. L. Anderson and B. A. Stone (1979). Form and function of arabinogalactans and arabinogalactan-proteins. *Phytochemistry* **18**(4): 521–540.

Clifton-Brown, J. C., W. C. Chiang and T. R. Hodkinson (2008). Miscanthus: Genetic resources and breeding potential to enhance bioenergy production. In *Genetic Improvement of Bioenergy Crops*, W. Vermerris (Ed.), pp. 273–294. New York: Springer Science.

Coenen, G. J., E. J. Bakx, R. P. Verhoef, H. A. Schols and A. G. J. Voragen (2007). Identification of the connecting linkage between homo- or xylogalacturonan and rhamnogalacturonan type I. *Carbohydr Polym* **70**(2): 224–235.

Coffman, A. M., Q. Li and L.-K. Ju (2014). Effect of natural and pretreated soybean hulls on enzyme production by *Trichoderma reesei*. *J Am Oil Chem Soc* **91**(8): 1331–1338.

Cohen, R., J. P. Lata, Y. Lee, C. H. Hernandez, N. Nishimura, C. B. Schaffer, C. Mukai et al. (2015). Use of tethered enzymes as a platform technology for rapid analyte detection. *PLoS One* **10**(11): e0142326.

Cohen, R., M. R. Suzuki and K. E. Hammel (2005). Processive endoglucanase active in crystalline cellulose hydrolysis by the brown rot basidiomycete *Gloeophyllum trabeum*. *Appl Environ Microbiol* **71**(5): 2412–2417.

Colquhoun, I. J., M. C. Ralet, J. F. Thibault, C. B. Faulds and G. Williamson (1994). Structure identification of feruloylated oligosaccharides from sugar-beet pulp by NMR spectroscopy. *Carbohyndr Res* **263**(2): 243–256.

Conde-Mejía, C., A. Jiménez-Gutiérrez and M. El-Halwagi (2012). A comparison of pretreatment methods for bioethanol production from lignocellulosic materials. *Process Saf Environ Prot* **90**(3): 189–202.

Cornet, P., J. Millet, P. Beguin and J.-P. Aubert (1983). Characterization of two Cel (cellulose degradation) genes of *Clostridium Thermocellum* coding for endoglucanases. *Nat Biotech* **1**(7): 589–594.

Cosgrove, D. J. (2001). Wall structure and wall loosening. A look backwards and forwards. *Plant Physiol* **125**(1): 131–134.

Coughlan, M. P., K. Hon-Nami, H. Hon-Nami, L. G. Ljungdahl, J. J. Paulin and W. E. Rigsby (1985). The cellulolytic enzyme complex of *Clostridium thermocellum* is very large. *Biochem Biophys Res Commun* **130**(2): 904–909.

Cousins, S. K. and R. M. Brown Jr (1995). Cellulose I microfibril assembly-computational molecular mechanic energy analysis favours bonding by van der Waals forcesas the initial step in crystallization. *Polymer* **36**(20): 3885–3888.

Crameri, A., S. A. Raillard, E. Bermudez and W. P. Stemmer (1998). DNA shuffling of a family of genes from diverse species accelerates directed evolution. *Nature* **391**(6664): 288–291.

Crawford, D. L., A. L. Pometto and R. L. Crawford (1983). Lignin degradation by *Streptomyces viridosporus*: Isolation and characterization of a new polymeric lignin degradation intermediate. *Appl Environ Microbiol* **45**(3): 898–904.

Crestini, C. and D. S. Argyropoulos (1998). The early oxidative biodegradation steps of residual kraft lignin models with laccase. *Bioorg Med Chem* **6**: 2161–2169.

Currie, M. A., J. J. Adams, F. Faucher, E. A. Bayer, Z. Jia and S. P. Smith (2012). Scaffoldin conformation and dynamics revealed by a ternary complex from the *Clostridium thermocellum* cellulosome. *J Biol Chem* **287**(32): 26953–26961.

Currie, M. A., K. Cameron, F. M. Dias, H. L. Spencer, E. A. Bayer, C. M. Fontes, S. P. Smith and Z. Jia (2013). Small angle X-ray scattering analysis of *Clostridium thermocellum* cellulosome N-terminal complexes reveals a highly dynamic structure. *J Biol Chem* **288**(11): 7978–7985.

Cybulska, I., G. Brudecki, K. Rosentrater, J. L. Julson and H. Lei (2012). Comparative study of organosolv lignin extracted from prairie cordgrass, switchgrass and corn stover. *Bioresour Technol* **118**: 30–36.

d'Acunzo, F., C. Galli, P. Gentili and F. Sergi (2006). Mechanistic and steric issues in the oxidation of phenolic and non-phenolic compounds by laccase or laccase-mediator systems. The case of bifunctional substrates. *New J Chem* **30**(4): 583–591.

da Costa Lopes, A. M., K. G. João, A. R. C. Morais, E. Bogel-Łukasik and R. Bogel-Łukasik (2013). Ionic liquids as a tool for lignocellulosic biomass fractionation. *Sustain Chem Process* **1**(3): 1–31.

Dadi, A. P., S. Varanasi and C. A. Schall (2006). Enhancement of cellulose saccharification kinetics using an ionic liquid pretreatment step. *Biotechnol Bioeng* **95**(5): 904–910.

Dale, B. E. and M. J. Moreira (1983). Freeze-explosion technique for increasing cellulose hydrolysis. *Biotechnology and Bioengineering Symposium*, 12 ed, pp. 31–43. John Wiley & Sons.

Daniel, G., J. Volc and E. Kubatova (1994). Pyranose oxidase, a major source of H₂O₂ during wood degradation by *Phanerochaete chrysosporium, Trametes versicolor,* and *Oudemansiella mucida. Appl Environ Microbiol* **60**: 2524–2532.

Das, N. N., S. C. Das and A. K. Mukherjee (1984). On the ester linkage between lignin and 4-O-methyl-D-glucurono-D-xylan in jute fiber (*Corchorus capsularis*). *Carbohydr Res* **127**(2): 345–348.

Dassa, B., I. Borovok, R. Lamed, B. Henrissat, P. Coutinho, C. L. Hemme, Y. Huang, J. Zhou and E. A. Bayer (2012). Genome-wide analysis of *Acetivibrio cellulolyticus* provides a blueprint of an elaborate cellulosome system. *BMC Genomics* **13**(1): 210.

Datta, S., B. Holmes, J. I. Park, Z. Chen, D. C. Dibble, M. Hadi, H. W. Blanch, B. A. Simmons and R. Sapra (2010). Ionic liquid tolerant hyperthermophilic cellulases for biomass pretreatment and hydrolysis. *Green Chem* **12**(2): 338–345.

Davidi, L., S. Morais, L. Artzi, D. Knop, Y. Hadar, Y. Arfi and E. A. Bayer (2016). Toward combined delignification and sacchrification of wheat straw by a laccase-containing designer cellulosome. *PNAS* **113**(39): 10854–10859.

Davids, T., M. Schmidt, D. Bottcher and U. T. Bornscheuer (2013). Strategies for the discovery and engineering of enzymes for biocatalysis. *Curr Opin Chem Biol* **17**(2): 215–220.

Davidsen, T., E. Beck, A. Ganapathy, R. Montgomery, N. Zafar, Q. Yang, R. Madupu, P. Goetz, K. Galinsky, O. White and G. Sutton (2010). The comprehensive microbial resource (CMR). *Nucleic Acids Res* **38**: D340–D345.

Dawson, D. H. (1972). History and organization of the maximum wood yield program. In *Intensive Plantation Culture,* J. H. Ohman (Ed.). St. Paul, MN: U.S. Department Agriculture, Forest Service, North Central Forest Experiment Station.

de Barros, R. R., S. Paredes Rde, T. Endo, E. P. Bon and S. H. Lee (2013). Association of wet disk milling and ozonolysis as pretreatment for enzymatic saccharification of sugarcane bagasse and straw. *Bioresour Technol* **136**: 288–294.

De Groeve, M. R., M. De Baere, L. Hoflack, T. Desmet, E. J. Vandamme and W. Soetaert (2009). Creating lactose phosphorylase enzymes by directed evolution of cellobiose phosphorylase. *Protein Eng Des Sel* **22**(7): 393–399.

de Jong, E., A. Higson, P. Walsh and M. Wellisch (2012). *Bio-based Chemicals: Value Added Products from Biorefineries.* Amsterdam, the Netherlands: IEA Bioenergy-Task42 Biorefinery, pp. 1–34.

De Mot, R. (1990). Conversion of starch by yeasts. In *Yeast Biotechnology and Biocatalysis,* H. Verchtert and R. De Mot (Eds.), pp. 163–222. Amsterdam, the Netherlands: Dekker.

De Vries, J. A., F. M. Rombouts, A. G. J. Voragen and W. Pilnik (1982). Enzymatic degradation of apple pectins. *Carbohydr Polym* **2**(1): 25–33.

de Vries, R. P. and J. Visser (2001). *Aspergillus* enzymes involved in degradation of plant cell wall polysaccharides. *Microbiol Mol Biol Rev* **65**(4): 497–522.

de Vrije, T., G. G. de Haas, G. B. Tan, E. R. P. Keijsers and P. A. M. Claassen (2002). Pretreatment of *Miscanthus* for hydrogen production by *Thermotoga* elfii. *Int J Hydrogen Energy* **27**(11–12): 1381–1390.

Debell, D. S., P. E. Heilman and D. V. J. Peabody (1972). *Potential Production of Black Cottonwood and Red Alder at Dense Spacings in the Pacific Northwest,* p. 2. Appleton, WI: Institute of Paper Chemistry.

Dekker, R. F. H. (1983). Bioconversion of hemicellulose: Aspects of hemicellulase production by *Trichoderma reesei* QM 9414 and enzymic saccharification of hemicellulose. *Biotechnol Bioeng* **25**(4): 1127–1146.

Delgado, L., A. T. Blanco, C. Huitron and G. Aguilar (1992). Pectin lyase from *Aspergillus* sp. CH-Y-1043. *Appl Microbiol Biotechnol* **39**(4): 515–519.

Delgenes, J. P., R. Moletta and J. M. Navarro (1990). Acid-hydrolysis of wheat straw and process considerations for ethanol fermentation by *Pichia-Stipitis* Y7124. *Process Biochem* **25**: 132–135.

Delucchi, M. A. (2010). Impacts of biofuels on climate change, water use, and land use. *Ann N Y Acad Sci* **1195**: 28–45.

Demain, A. L., M. Newcomb and J. H. D. Wu (2005). Cellulase, clostridia, and ethanol. *Microbiol Mol Biol Rev* **69**(1): 124–154.

Dence, C. W. (1992). The determination of lignin. In *Methods in Lignin Chemistry*, S. Y. Lin and C. W. Deuce (Eds.), pp. 35–57. Berlin, Germany: Springer.

Deng, L., Y. Mori, J. Sermsathanaswadi, W. Apiwantanapiwat and A. Kosugi (2015). Cellulose hydrolysis ability of a *Clostridium thermocellum* cellulosome containing small-size scaffolding protein CipA. *J Biotechnol* **212**: 144–152.

Derksen, J., G. J. Janssen, M. Wolters-Arts, I. Lichtscheidl, W. Adlassnig, M. Ovecka, F. Doris and M. Steer (2011). Wall architecture with high porosity is established at the tip and maintained in growing pollen tubes of *Nicotiana tabacum*. *Plant J* **68**(3): 495–506.

Desjarlais, J. R. and N. D. Clarke (1998). Computer search algorithms in protein modification and design. *Curr Opin Struct Biol* **8**(4): 471–475.

Despotovic, D., L. Vojcic, R. Prodanovic, R. Martinez, K. H. Maurer and U. Schwaneberg (2012). Fluorescent assay for directed evolution of perhydrolases. *J Biomol Screen* **17**(6): 796–805.

Detroy, R. W., R. L. Cunningham, R. J. Bothast, M. O. Bagby and A. Herman (1982). Bioconversion of wheat straw cellulose/hemicellulose to ethanol by *Saccharomyces uvarum* and *Pachysolen tannophilus*. *Biotechnol Bioeng* **24**(5): 1105–1113.

Dhillon, A. and S. Khanna (2000). Production of a thermostable alkali-tolerant xylanase from *Bacillus circulans* AB 16 grown on wheat straw. *World J Microbiol Biotechnol* **16**: 325–327.

Dias, M. O., M. P. da Cunha, R. Maciel Filho, A. Bonomi, C. D. Jesus and C. E. Rossell (2011). Simulation of integrated first and second generation bioethanol production from sugarcane: Comparison between different biomass pretreatment methods. *J Ind Microbiol Biotechnol* **38**(8): 955–966.

Dias, M. O. S., A. V. Ensinas, S. A. Nebra, R. Maciel Filho, C. E. V. Rossell and M. R. W. Maciel (2009). Production of bioethanol and other bio-based materials from sugarcane bagasse: Integration to conventional bioethanol production process. *Chem Eng Res Des* **87**(9): 1206–1216.

Dibble, D. C., C. Li, L. Sun, A. George, A. Cheng, O. P. Cetinkol, P. Benke, B. M. Holmes, S. Singh and B. A. Simmons (2011). A facile method for the recovery of ionic liquid and lignin from biomass pretreatment. *Green Chem* **13**(11): 3255–3264.

Dick-Perez, M., Y. Zhang, J. Hayes, A. Salazar, O. A. Zabotina and M. Hong (2011). Structure and interactions of plant cell-wall polysaccharides by two- and three-dimensional magic-angle-spinning solid-state NMR. *Biochemistry* **50**(6): 989–1000.

Diedericks, D., E. van Rensburg and J. F. Görgens (2012). Fractionation of sugarcane bagasse using a combined process of dilute acid and ionic liquid treatments. *Appl Biochem Biotechnol* **167**: 1921–1937.

Dien, B. S., L. B. Iten and C. D. Skory (2005). Converting herbaceous energy crops to bioethanol, a review with emphasis on pretreatment processes. In *Handbook of Industrial Biocatalysis*, C. T. Hou (Ed.), pp. 1–11. Boca Raton, FL: Taylor & Francis Group.

Dien, B. S., G. Sarath, J. Pedersen, K. Vogel, H.-J. G. Jung, S. Sattler, M. D. Casler, R. B. Michell and M. A. Cotta (2008). Energy crops for ethanol: A processing perspective. *Proceedings of the 5th International Crop Science Congress*, April 13–18, Jeju Island, Korea.

Digman, M. F., K. J. Shinners, M. D. Casler, B. S. Dien, R. D. Hatfield, H. J. Jung, R. E. Muck and P. J. Weimer (2010). Optimizing on-farm pretreatment of perennial grasses for fuel ethanol production. *Bioresour Technol* **101**(14): 5305–5314.

Dimarogona, M., E. Topakas and P. Christakopoulos (2012). Cellulose degradation by oxidative enzymes. *Comput Struct Biotechnol J* **2**: 1–8.

Din, N., H. G. Damude, N. R. Gilkes, R. C. Miller Jr., R. A. Warren and D. G. Kilburn (1994). C1-Cx revisited: Intramolecular synergism in a cellulase. *Proc Natl Acad Sci USA* **91**(24): 11383–11387.

Ding, S. Y., E. A. Bayer, D. Steiner, Y. Shoham and R. Lamed (1999). A novel cellulosomal scaffoldin from *Acetivibrio cellulolyticus* that contains a family 9 glycosyl hydrolase. *J Bacteriol* **181**(21): 6720–6729.

Ding, S. Y., Y. S. Liu, Y. Zeng, M. E. Himmel, J. O. Baker and E. A. Bayer (2012). How does plant cell wall nanoscale architecture correlate with enzymatic digestibility? *Science* **338**(6110): 1055–1060.

Divne, C., J. Stahlberg, T. Reinikainen, L. Ruohonen, G. Pettersson, J. K. Knowles, T. T. Teeri and T. A. Jones (1994). The three-dimensional crystal structure of the catalytic core of cellobiohydrolase I from *Trichoderma reesei*. *Science* **265**(5171): 524–528.

Docherty, K. M. and J. C. F. Kulpa (2005). Toxicity and antimicrobial activity of imidazolium and pyridinium ionic liquids. *Green Chem* **7**(4): 185–189.

Doherty, T. V., M. Mora-Pale, S. E. Foley, R. J. Linhardt and J. S. Dordick (2010). Ionic liquid solvent properties as predictors of lignocellulose pretreatment efficacy. *Green Chem* **12**: 1967–1975.

Domanska, U. and R. Bogel-Lukasik (2005). Physicochemical properties and solubility of alkyl-(2-hydroxyethyl)-dimethylammonium bromide. *J Phys Chem B* **109**(24): 12124–12132.

Dominguez-Faus, R., S. E. Powers, J. G. Burken and P. J. Alvarez (2009). The water footprint of biofuels: A drink or drive issue? *Environ Sci Technol* **43**(9): 3005–3010.

Doucet, N. (2011). Can enzyme engineering benefit from the modulation of protein motions? Lessons learned from NMR relaxation dispersion experiments. *Protein Pept Lett* **18**(4): 336–343.

Drevland, R. M., J. W. Cunha, H. Tran, J. Sustarich, P. D. Adams, A. K. Singh, B. A. Simmons and K. L. Sale (2014). Directed evolution of a beta-glucosidase for ionic liquid tolerance. *Paper Presented at the 36th Symposium on Biotechnology for Fuels and Chemicals*, April 28–May 1, Clearwater Beach, FL.

Drummond, D. A., J. D. Bloom, C. Adami, C. O. Wilke and F. H. Arnold (2005). Why highly expressed proteins evolve slowly. *Proc Natl Acad Sci USA* **102**(40): 14338–14343.

Du, B., L. N. Sharma, C. Becker, S.-F. Chen, R. A. Mowery, G. P. van Walsum and C. K. Chambliss (2010). Effect of varying feedstock–pretreatment chemistry combinations on the formation and accumulation of potentially inhibitory degradation products in biomass hydrolysates. *Biotechnol Bioeng* **107**(3): 430–440.

Du, W., Q. Wang, J.-K. Wang and J.-X. Liu (2014). Enhancing catalytic activity of a xylanase retrieved from a fosmid library of rumen microbiota in Hu sheep by directed evolution. *J Anim Vet Adv* **13**(8): 538–544.

Dunlap, C. E. and L. C. Chiang (1980). Cellulose degradation-a common link. In *Utilization and Recycle of Agricultural Wastes and Residues*, M. L. Shuler (Ed.), pp. 19–65. Boca Raton, FL: CRC Press.

Ďuranová, M., J. Hirsch, K. Kolenova and P. Biely (2009). Fungal glucuronoyl esterases and substrate uronic acid recognition. *Biosci Biotechnol Biochem* **73**(11): 2483–2487.

Dwivedi, U. N., P. Singh, V. P. Pandey and A. Kumar (2011). Structure–function relationship among bacterial, fungal and plant laccases. *J Mol Catal B: Enzym* **68**(2): 117–128.

Dworschack, R. G. and L. J. Wickerham (1961). Production of extracellular and total Invertase by *Candida utilis, Saccharomyces cerevisiae*, and other yeasts. *Appl Environ Microbiol* **9**(4): 291–294.

Edstrom, R. D. and H. J. Phaff (1963). Purification and certain properties of pectin *trans*-eliminase from *Aspergihs fonsecaeus*. *J Biol Chem* **239**: 2403–2408.

Edwards, M. C. and J. Doran-Peterson (2012). Pectin-rich biomass as feedstock for fuel ethanol production. *Appl Microbiol Biotechnol* **95**(3): 565–575.

Edwards, M. C., E. D. Henriksen, L. P. Yomano, B. C. Gardner, L. N. Sharma, L. O. Ingram and J. Doran Peterson (2011). Addition of genes for cellobiase and pectinolytic activity in *Escherichia coli* for fuel ethanol production from pectin-rich lignocellulosic biomass. *Appl Environ Microbiol* **77**(15): 5184–5191.

Egüés, I., C. Sanchez, I. Mondragon and J. Labidi (2012). Effect of alkaline and autohydrolysis processes on the purity of obtained hemicelluloses from corn stalks. *Bioresour Technol* **103**(1): 239–248.

Emmel, A., A. L. Mathias, F. Wypych and L. P. Ramos (2003). Fractionation of Eucalyptus grandis chips by dilute acid-catalysed steam explosion. *Bioresour Technol* **86**: 105–115.

Energy Information Administration (2004). Annual Energy Outlook 2004 with projections to 2025. U.S. Energy Information Administration.

Eneyskaya, E. V., H. Brumer III, L. V. Backinowsky, D. R. Ivanen, A. A. Kulminskaya, K. A. Shabalin and K. N. Neustroev (2003). Enzymatic synthesis of β-xylanase substrates: Transglycosylation reactions of the β-xylosidase from *Aspergillus* sp. *Carbohydr Res* **338**(4): 313–325.

Eneyskaya, E. V., D. R. Ivanen, K. S. Bobrov, L. S. Isaeva-Ivanova, K. A. Shabalin, A. N. Savel'ev, A. M. Golubev and A. A. Kulminskaya (2007). Biochemical and kinetic analysis of the GH3 family β-xylosidase from *Aspergillus awamori* X-100. *Arch Biochem Biophys* **457**(2): 225–234.

Engel, P., R. Mladenov, H. Wulfhorst, G. Jager and A. C. Spiess (2010). Point by point analysis: How ionic liquid affects the enzymatic hydrolysis of native and modified cellulose. *Green Chem* **12**(11): 1959–1966.

Ensinas, A. V., M. Modesto, S. A. Nebra and L. Serra (2009). Reduction of irreversibility generation in sugar and ethanol production from sugarcane. *Energy* **34**(5): 680–688.

Eriksen, D. T., J. Lian and H. Zhao (2014). Protein design for pathway engineering. *J Struct Biol* **185**(2): 234–242.

Eriksson, K.-E. (1981). Cellulases of fungi. In *Trends in the Biology of Fermentations*, A. Hollaender (Ed.), pp. 19–32. New York: Plenum Press.

Eriksson, K. E., B. Pettersson and U. Westermark (1974). Oxidation: An important enzyme reaction in fungal degradation of cellulose. *FEBS Lett* **49**(2): 282–285.

Escalante, A., A. Gonçalves, A. Bodin, A. Stepan, C. Sandström, G. Toriz and P. Gatenholm (2012). Flexible oxygen barrier films from spruce xylan. *Carbohydr Polym* **87**(4): 2381–2387.

Escovar-Kousen, J. M., D. Wilson and D. Irwin (2004). Integration of computer modeling and initial studies of site-directed mutagenesis to improve cellulase activity on Cel9A from *Thermobifida fusca*. *Appl Biochem Biotechnol* **113– 116**: 287–297.

Evans, C. S., M. V. Dutton, F. Guillén and R. G. Veness (1994). Enzymes and small molecular mass agents involved with lignocellulose degradation. *FEMS Microbiol Rev* **13**(2–3): 235–239.

Fackler, K., C. Gradinger, B. Hinterstoisser, K. Messner and M. Schwanninger (2006). Lignin degradation by white rot fungi on spruce wood shavings during short-time solid-state fermentations monitored by near infrared spectroscopy. *Enzyme Microbial Technol* **39**(7): 1476–1483.

Fägerstam, L. G. and L. G. Pettersson (1980). The 1,4-β-glucan cellobiohydrolases of *Trichoderma reesei* QM 9414: A new type of cellulolytic synergism. *FEBS Lett* **119**(1): 97–100.

Fagerstedt, K. V., E. M. Kukkola, V. V. Koistinen, J. Takahashi and K. Marjamaa (2010). Cell wall lignin is polymerised by class III secretable plant peroxidases in Norway spruce. *J Integr Plant Biol* **52**(2): 186–194.

Fan, L. T., M. M. Gharpuray and Y. H. Lee (1987). Cellulose hydrolysis. In *Biotechnology Monographs*, S. Aiba, L.T. Fan, A. Fiechter, J. de Klein and K. Schügerl (Eds.), Vol. 3, p. 57. Berlin, Germany: Springer.

Fan, L. T., Y.-H. Lee and M. M. Gharpuray (1982). The nature of lignocellulosics and their pretreatments for enzymatic hydrolysis. In *Microbial Reactions*, T. Scheper, S. Belkin, T. Bley, J. Bohlmann, M. B. Gu, W. S. Hu, B. Mattiasson et al. (Eds.), Vol. 23, pp. 157–187. Berlin, Germany: Springer.

Fang, Z., P. Zhou, F. Chang, Q. Yin, W. Fang, J. Yuan, X. Zhang and Y. Xiao (2014). Structure-based rational design to enhance the solubility and thermostability of a bacterial laccase Lac15. *PLoS One* **9**(7): e102423.

Fanutti, C., T. Ponyi, G. W. Black, G. P. Hazlewood and H. J. Gilbert (1995). The conserved noncatalytic 40-residue sequence in cellulases and hemicellulases from anaerobic fungi functions as a protein docking domain. *J Biol Chem* **270**(49): 29314–29322.

Fauth, U., M. P. Romaniec, T. Kobayashi and A. L. Demain (1991). Purification and characterization of endoglucanase Ss from *Clostridium thermocellum*. *Biochem J* **279**(1): 67–70.

Fengel, D. and G. Wegener (1984). *Wood: Chemistry, Ultrastructure, Reactions*. Berlin, Germany: Walter de Gruyter Publishers.

Fernandez, E. C., A. M. Palijon, W. Liese, F. L. Esguerra and R. J. Murphy (2003). Silviculture and managemenet of newly established Bamboo plantation. *XII World Forestry Congress*, Quebec City, Canada.

Fernandez-Alvaro, E., R. Snajdrova, H. Jochens, T. Davids, D. Bottcher and U. T. Bornscheuer (2011). A combination of in vivo selection and cell sorting for the identification of enantioselective biocatalysts. *Angew Chem Int Ed Engl* **50**(37): 8584–8587.

Ferre, H., A. Broberg, J. O. Duus and K. K. Thomsen (2000). A novel type of arabinoxylan arabinofuranohydrolase isolated from germinated barley analysis of substrate preference and specificity by nano-probe NMR. *Eur J Biochem* **267**(22): 6633–6641.

Ferrés, I., V. Amarelle, F. Noya and E. Fabiano (2015). Identification of Antarctic culturable bacteria able to produce diverse enzymes of potential biotechnological interest *Adv Polar Sci* **26**(1): 71–79.

Fiberright, LLC. (2015). Retrieved May 5, 2015, from http://fiberight.com/.

Fierobe, H. P., E. A. Bayer, C. Tardif, M. Czjzek, A. Mechaly, A. Belaich, R. Lamed, Y. Shoham and J. P. Belaich (2002). Degradation of cellulose substrates by cellulosome chimeras. Substrate targeting versus proximity of enzyme components. *J Biol Chem* **277**(51): 49621–49630.

Fierobe, H. P., S. Pagès, A. Bélaïch, S. Champ, D. Lexa and J. P. Bélaïch (1999). Cellulosome from *Clostridium cellulolyticum*: Molecular study of the Dockerin/Cohesin interaction. *Biochemistry* **38**(39): 12822–12832.

Fierobe, H.-P., A. Mechaly, C. Tardif, A. Belaich, R. Lamed, Y. Shoham, J.-P. Belaich and E. A. Bayer (2001). Design and production of active cellulosome chimeras. Selective incorporation of dockerin-containing enzymes into defined functional complexes. *J Biol Chem* **276**(24): 21257–21261.

Fierobe, H.-P., F. Mingardon, A. Mechaly, A. Bélaïch, M. T. Rincon, S. Pagès, R. Lamed, C. Tardif, J.-P. Bélaïch and E. A. Bayer (2005). Action of designer cellulosomes on homogeneous versus complex substrates: Controlled incorporation of three distinct enzymes into a defined trifunctional scaffoldin. *J Biol Chem* **280**(16): 16325–16334.

Fillingham, I. J., P. A. Kroon, G. Williamson, H. J. Gilbert and G. P. Hazlewood (1999). A modular cinnamoyl ester hydrolase from the anaerobic fungus *Piromyces equi* acts synergistically with xylanase and is part of a multiprotein cellulose-binding cellulase-hemicellulase complex. *Biochem J* **343**(1): 215–224.

Fincher, G. B. and B. A. Stone (1981). Metabolism of noncellulosic polysaccharides. In *Plant Carbohydrates II*, W. Tanner and F. A. Loewus (Eds.), pp. 68–132. Berlin, Germany: Springer.

Flournoy, D. S., J. A. Paul, T. K. Kirk and T. L. Highley (1993). Changes in the size and volume of pores in sweetgum wood during simultaneous rot by *Phanerochaete chrysosporium* burds. *Holzforschung* **47**: 297–301.

Fonseca-Maldonado, R., D. S. Vieira, J. S. Alponti, E. Bonneil, P. Thibault and R. J. Ward (2013). Engineering the pattern of protein glycosylation modulates the thermostability of a GH11 xylanase. *J Biol Chem* **288**(35): 25522–25534.

Fontes, C. M. G. A. and H. J. Gilbert (2010). Cellulosomes: Highly efficient nanomachines designed to deconstruct plant cell wall complex carbohydrates. *Annu Rev Biochem* **79**(1): 655–681.

Fontes, C. M., G. P. Hazlewood, E. Morag, J. Hall, B. H. Hirst and H. J. Gilbert (1995). Evidence for a general role for non-catalytic thermostabilizing domains in xylanases from thermophilic bacteria. *Biochem J* **307**(Pt 1): 151–158.

Food & Agriculture Organization of the United Nations (1992). Gross chemical composition. In *Maize in Human Nutrition*, p. 25. Rome, Italy: David Lubin Memorial Library.

Food & Agriculture Organization of the United Nations (2009). Pectins. *FAO JECFA Monographs 7*.

Food and Agriculture Organization of the United Nations (2008). The state of food and agriculture. Biofuels: Prospects, risks and opportunities. Rome, Italy.

Foreman, P. K., D. Brown, L. Dankmeyer, R. Dean, S. Diener, N. S. Dunn-Coleman, F. Goedegebuur et al. (2003). Transcriptional regulation of biomass-degrading enzymes in the filamentous fungus *Trichoderma reesei*. *J Biol Chem* **278**(34): 31988–31997.

Forsberg, Z., G. Vaaje-Kolstad, B. Westereng, A. C. Bunaes, Y. Stenstrom, A. MacKenzie, M. Sorlie, S. J. Horn and V. G. Eijsink (2011). Cleavage of cellulose by a CBM33 protein. *Protein Sci* **20**(9): 1479–1483.

Fort, D. A., R. C. Remsing, R. P. Swatloski, P. Moyna, G. Moyna and R. D. Rogers (2007). Can ionic liquids dissolve wood? Processing and analysis of lignocellulosic materials with 1-n-butyl-3-methylimidazolium chloride. *Green Chem* **9**(1): 63–69.

Fowler, T. and R. D. Brown (1992). The bgI1 gene encoding extracellular β-glucosidase from *Trichoderma reesei* is required for rapid induction of the cellulase complex. *Mol Microbiol* **6**(21): 3225–3235.

Fox, R. J., S. C. Davis, E. C. Mundorff, L. M. Newman, V. Gavrilovic, S. K. Ma, L. M. Chung et al. (2007). Improving catalytic function by ProSAR-driven enzyme evolution. *Nat Biotechnol* **25**(3): 338–344.

Franke, W. and M. Deffner (1939). Zur Kenntnis der sog. Glucose-oxydase. II. *Liebigs Ann* **541**(1): 117–150.

Franke, W. and F. Lorenz (1937). Zur Kenntnis der sog. Glucose-oxydase. I. *Liebigs Ann* **532**(1): 1–28.

Fratzl, P., I. Burgert and H. S. Gupta (2004). On the role of interface polymers for the mechanics of natural polymeric composites. *Phys Chem Chem Phys* **6**(20): 5575–5579.

Freer, S. N. (1993). Kinetic characterization of a β-glucosidase from a yeast, Candida wickerhamii. *J Biol Chem* **268**(13): 9337–9342.

Fries, M., J. Ihrig, K. Brocklehurst, V. E. Shevchik and R. W. Pickersgill (2007). Molecular basis of the activity of the phytopathogen pectin methylesterase. *EMBO J* **26**(17): 3879–3887.

Fry, S. C. (1979). Phenolic components of the primary cell wall and their possible role in the hormonal regulation of growth. *Planta* **146**(3): 343–351.

Fry, S. C. (1983). Feruloylated pectins from the primary cell wall: Their structure and possible functions. *Planta* **157**(2): 111–123.

Fu, J. and A. Mort (1997). Progress towards identifying a covalent cross-link between xyloglucan and rhamnogalacturonan in cotton cell walls (Abstr). *Plant Physiol* **114S**: 83.

Fujii, T., G. Yu, A. Matsushika, A. Kurita, S. Yano, K. Murakami and S. Sawayama (2011). Ethanol production from xylo-oligosaccharides by xylose-fermenting *Saccharomyces cerevisiae* expressing β-xylosidase. *Biosci Biotechnol Biochem* **75**(6): 1140–1146.

Fujino, T., P. Béguin and J. P. Aubert (1992). Cloning of a *Clostridium thermocellum* DNA fragment encoding polypeptides that bind the catalytic components of the cellulosome. *FEMS Microbiol Lett* **73**(1–2): 165–170.

Furlan, F. F., C. B. B. Costa, G. D. C. Fonseca, R. D. P. Soares, A. R. Secchi, A. J. G. D. Cruz and R. D. C. Giordano (2012). Assessing the production of first and second generation bioethanol from sugarcane through the integration of global optimization and process detailed modeling. *Comput Chem Eng* **43**(10): 1–9.

Furukawa, T., F. O. Bello and L. Horsfall (2014). Microbial enzyme systems for lignin degradation and their transcriptional regulation. *Front Biol* **9**(6): 448–471.

Fushinobu, S., M. Hidaka, Y. Honda, T. Wakagi, H. Shoun and M. Kitaoka (2005). Structural basis for the specificity of the reducing end xylose-releasing exo-oligoxylanase from *Bacillus halodurans* C-125. *J Biol Chem* **280**(17): 17180–17186.

Gabrielii, I., P. Gatenholm, W. G. Glasser, R. K. Jain and L. Kenne (2000). Separation, characterization and hydrogel-formation of hemicellulose from aspen wood. *Carbohydr Polym* **43**(4): 367–374.

Galante, Y. M., A. De Conti and R. Monteverdi (1998). Application of Trichoderma enzymes in the textile industry. In *Trichoderma and Gliocladium*, G. E. Harman and C. P. Kubicek (Eds.), Vol. 1, pp. 311–326. London, UK: Taylor & Francis Group.

Galkin, S., T. Vares, M. Kalsi and A. Hatakka (1998). Production of organic acids by different white-rot fungias detected using capillary zone electrophoresis. *Biotechnol Technol* **12**: 267–271.

Ganju, R. K., P. J. Vithayathil and S. K. Murthy (1989). Purification and characterization of two xylanases from *Chaetomium thermophile* var. *coprophile*. *Can J Microbiol* **35**(9): 836–842.

Gao, D., N. Uppugundla, S. P. Chundawat, X. Yu, S. Hermanson, K. Gowda, P. Brumm, D. Mead, V. Balan and B. E. Dale (2011). Hemicellulases and auxiliary enzymes for improved conversion of lignocellulosic biomass to monosaccharides. *Biotechnol Biofuels* **4**(1): 5.

Garcia-Alvarez, B., R. Melero, F. M. Dias, J. A. Prates, C. M. Fontes, S. P. Smith, M. J. Romao, A. L. Carvalho and O. Llorca (2011). Molecular architecture and structural transitions of a *Clostridium thermocellum* mini-cellulosome. *J Mol Biol* **407**(4): 571–580.

García-Aparicio, M., W. Parawira, E. Van Rensburg, D. Diedericks, M. Galbe, C. Rosslander, G. Zacchi and J. Görgens (2011). Evaluation of steam-treated giant bamboo for production of fermentable sugars. *Biotechnol Prog* **27**(3): 641–649.

García-Cubero, M. A., G. González-Benito, I. Indacoechea, M. Coca and S. Bolado (2009). Effect of ozonolysis pretreatment on enzymatic digestibility of wheat and rye straw. *Bioresour Technol* **100**(4): 1608–1613.

Garcia-Martinez, D. V., A. Shinmyo, A. Madia and A. L. Demain (1980). Studies on cellulase production by *Clostridium thermocellum*. *Eur J Appl Microbiol Biotechnol* **9**(3): 189–197.

Garcia-Ruiz, E., D. Gonzalez-Perez, F. J. Ruiz-Dueñas, A. T. Martínez and M. Alcalde (2012). Directed evolution of a temperature-, peroxide- and alkaline pH-tolerant versatile peroxidase. *Biochem J* **441**(1): 487–498.

Gardner, R. D., K. E. Cooksey, F. Mus, R. Macur, K. Moll, E. Eustance, R. P. Carlson, R. Gerlach, M. W. Fields and B. M. Peyton (2012). Use of sodium biocarbonate to stimulate triacylglyerol accumulation in the chlorophyte *Scenedesmus* sp. and the diatom *Phaeodactylum tricornutum*. *J Appl Phycol* **24**: 1311–1320.

Garmakhany, A. D., M. Kashaninejad, M. Aalami, Y. Maghsoudlou, M. Khomieria and L. G. Tabil (2013). Enhanced biomass delignification and enzymatic saccharification of canola straw by steam-explosion pretreatment. *J Sci Food Agric* **94**(8): 1607–1613.

Garrote, G., H. Dominguez and J. C. Parajo (2002). Autohydrolysis of corncob: Study of non-isothermal operation for xylooligosaccharide production. *J Food Eng* **52**(3): 211–218.

Gasparic, A., J. Martin, A. S. Daniel and H. J. Flint (1995). A xylan hydrolase gene cluster in *Prevotella ruminicola* B(1)4: Sequence relationships, synergistic interactions, and oxygen sensitivity of a novel enzyme with exoxylanase and beta-(1,4)-xylosidase activities. *Appl Environ Microbiol* **61**(8): 2958–2964.

Ge, X., D. M. Burner, J. Xu, G. C. Phillips and G. Sivakumar (2011). Bioethanol production from dedicated energy crops and residues in Arkansas, USA. *Biotechnol J* **6**(1): 66–73.

Ge, X., V. S. Green, N. Zhang, G. Sivakumar and J. Xu (2012). Eastern gamagrass as an alternative cellulosic feedstock for bioethanol production. *Process Biochem* **47**(2): 335–339.

Gee, M., R. M. Reeve and R. M. Mcready (1959). Measurement of plant pectic substances, reaction of hydroxylamine with pectinic acids. Chemical studies and histochemical estimation of the degree of esterification of pectic substances in fruit. *J Agric Food Chem* **7**: 34–38.

Gerasimowicz, W. V., K. B. Hicks and P. E. Pfeffer (1984). Evidence for the existence of associated lignin-carbohydrate polymers as revealed by carbon-13 CPMAS solid-state NMR spectroscopy. *Macromolecules* **17**(12): 2597–2603.

Gerlt, J. A. and P. C. Babbitt (2009). Enzyme (re)design: Lessons from natural evolution and computation. *Curr Opin Chem Biol* **13**(1): 10–18.

Gerngross, U. T., M. P. Romaniec, T. Kobayashi, N. S. Huskisson and A. L. Demain (1993). Sequencing of a *Clostridium thermocellum* gene (cipA) encoding the cellulosomal SL-protein reveals an unusual degree of internal homology. *Mol Microbiol* **8**(2): 325–334.

Gerwig, G. J., P. de Waard, J. P. Kamerling, J. F. G. Vliegenthart, E. Morgenstern, R. Lamed and E. A. Bayer (1989). Novel O-linked carbohydrate chains in the cellulase complex (cellulosome) of *Clostridium thermocellum*. 3-O-Methyl-N-acetylglucosamine as a constituent of a glycoprotein. *J Biol Chem* **264**(2): 1027–1035.

Gerwig, G. J., J. P. Kamerling, J. F. Vliegenthart, E. Morag, R. Lamed and E. A. Bayer (1991). Primary structure of O-linked carbohydrate chains in the cellulosome of different *Clostridium thermocellum* strains. *Eur J Biochem* **196**(1): 115–122.

Gerwig, G. J., J. P. Kamerling, J. F. G. Vliegentthart, E. Morag, R. Lamed and E. A. Bayer (1992). Novel oligosaccharide constituents of the cellulase complex of *Bacteroides cellulosolvens*. *Eur J Biochem* **205**(2): 799–808.

Gerwig, G. J., J. P. Kamerling, J. F. Vliegenthart, E. Morag, R. Lamed and E. A. Bayer (1993). The nature of the carbohydrate-peptide linkage region in glycoproteins from the cellulosomes of *Clostridium thermocellum* and *Bacteroides cellulosolvens*. *J Biol Chem* **268**(36): 26956–26960.

Ghedalia, D. B. and J. Miron (1981). The effect of combined chemical and enzyme treatments on the saccharification and in vitro digestion rate of wheat straw. *Biotechnol Bioeng* **23**(4): 823–831.

Gilad, R., L. Rabinovich, S. Yaron, E. A. Bayer, R. Lamed, H. J. Gilbert and Y. Shoham (2003). CelI, a noncellulosomal family 9 enzyme from *Clostridium thermocellum*, is a processive endoglucanase that degrades crystalline cellulose. *J Bacteriol* **185**(2): 391–398.

Gilbert, H. J. (2007). Cellulosomes: Microbial nanomachines that display plasticity in quaternary structure. *Mol Microb* **63**(6): 1568–1576.

Gilbert, H. J., J. P. Knox and A. B. Boraston (2013). Advances in understanding the molecular basis of plant cell wall polysaccharide recognition by carbohydrate-binding modules. *Curr Opin Struct Biol* **23**(5): 669–677.

Gilkes, N. R., M. L. Langsford, D. G. Kilburn, J. Miller, Robert C. and R. A. J. Warren (1984). Mode of action and substrate specificities of cellulases from cloned bacterial genes. *J Biol Chem* **259**(16): 10455–10459.

Gilkes, N. R., R. A. Warren, R. C. J. Miller and D. G. Kilburn (1988). Precise excision of the cellulose binding domains from two *Cellulomonas fimi* cellulases by a homologous protease and the effect on catalysis. *J Biol Chem* **263**(21): 10401–10407.

Gírio, F. M., C. Fonseca, F. Carvalheiro, L. C. Duarte, S. Marques and R. Bogel-Łukasik (2010). Hemicelluloses for fuel ethanol: A review. *Bioresour Technol* **101**: 4775–4800.

Glenn, J. K., L. Akileswaran and M. H. Gold (1986). Mn(II) oxidation is the principal function of the extracellular Mn-peroxidase from *Phanerochaete chrysosporium*. *Arch Biochem Biophys* **251**(2): 688–696.

Glenn, J. K. and M. H. Gold (1985). Purification and characterization of an extracellular Mn(II)-dependent peroxidase from the lignin-degrading basidiomycete, *Phanerochaete chrysosporium*. *Arch Biochem Biophys* **242**(2): 329–341.

Glieder, A., E. T. Farinas and F. H. Arnold (2002). Laboratory evolution of a soluble, self-sufficient, highly active alkane hydroxylase. *Nat Biotechnol* **20**(11): 1135–1139.

Glumoff, T., P. J. Harvey, S. Molinari, M. Goble, G. Frank, J. M. Palmer, J. D. G. Smit and M. S. A. Leisola (1990). Lignin peroxidase from *Phanerochaete chrysosporium*. *Eur J Biochem* **187**(3): 515–520.

Goff, A. L., C. M. G. C. Renard, E. Bonnin and J.-F. Thibault (2001). Extraction, purification and chemical characterisation of xylogalacturonans from pea hulls. *Carbohydr Polym* **45**(4): 325–334.

Goldberg, R. N. (1975). Thermodynamics of hexokinase-catalyzed reactions. *Biophys Chem* **3**(3): 192–205.

Goldstein, I. S. (1981). *Organic Chemicals from Biomass*. Boca Raton, FL: CRC Press.

Gomez-Toribio, V., A. T. Martinez, M. J. Martinez and F. Guillen (2001). Oxidation of hydroquinones by the versatile ligninolytic peroxidase from *Pleurotus eryngii*. H_2O_2 generation and the influence of Mn2+. *Eur J Biochem* **268**(17): 4787–4793.

González-Blasco, G., J. Sanz-Aparicio, B. González, J. A. Hermoso and J. Polaina (2000). Directed evolution of β-Glucosidase a from *Paenibacillus polymyxa* to thermal resistance. *J Biol Chem* **275**(18): 13708–13712.

Gorbacheva, I. V. and N. A. Rodionova (1977). Studies on xylan-degrading enzymes. II. Action pattern of endo-1,4-β-xylanase from *Aspergillus niger* STR. 14 on xylan and xylooligosaccharides. *Biochim Biophys Acta* **484**: 94–102.

Gosh, B. K. and A. Gosh (1992). Degradation of cellulose by fungal cellulase. In *Microbial Degradation of Natural Products*, G. Winkelmann (Ed.), pp. 84–126. New York: VCH Publishers.

Gosselink, R. J. A., M. H. B. Snijder, A. Kranenbarg, E. R. P. Keijsers, E. de Jong and L. L. Stigsson (2004). Characterisation and application of NovaFiber lignin. *Ind Crops Prod* **20**(2): 191–203.

Goyal, G., S.-L. Tsai, B. Madan, N. A. DaSilva and W. Chen (2011). Simultaneous cell growth and ethanol production from cellulose by an engineered yeast consortium displaying a functional mini-cellulosome. *Microb Cell Fact* **10**(89): 1–8.

Gräbnitz, F., K. P. Rücknagel, M. Seiß and W. L. Staudenhauer (1989). Nucleotide sequence of the *Clostridium thermocellum* bglB gene encoding thermostable β-glucosidase B: Homology to fungal β-glucosidases. *Mol Gen Genet* **217**(1): 70–76.

Gräbnitz, F., M. Seiss, K. P. Rücknagel and W. L. Staudenbauer (1991). Structure of the β-glucosidase gene bglA of *Clostridium thermocellum*. *Eur J Biochem* **200**(2): 301–309.

Gracheck, S. J., D. B. Rivers, L. C. Woodford, K. E. Giddings and G. H. Emert (1981). Pretreatment of lignocellulosics to support cellulase production using *Trichoderma reesei* QM9414. *Biotechnol Bioeng Symp* **11**: 47–65.

Graham, R. L., R. Nelson, J. Sheehan, R. D. Perlack and L. L. Wright (2007). Current and potential U.S. corn stover supplies. *Agron J* **99**(1): 1–11.

Grépinet, O. and P. Béguin (1986). Sequence of the cellulase gene of *Clostridium thermocellum* coding for endoglucanase B. *Nucl Acids Res* **14**(4): 1791–1799.

Grépinet, O., M. C. Chebrou and P. Béguin (1988). Purification of *Clostridium thermocellum* xylanase Z expressed in *Escherichia coli* and identification of the corresponding product in the culture medium of *C. thermocellum*. *J Bacteriol* **170**(10): 4576–4581.

Griffiths, J. S., M. Cheriyan, J. B. Corbell, L. Pocivavsek, C. A. Fierke and E. J. Toone (2004). A bacterial selection for the directed evolution of pyruvate aldolases. *Bioorg Med Chem* **12**(15): 4067–4074.

Grishutin, S. G., A. V. Gusakov, A. V. Markov, B. B. Ustinov, M. V. Semenova and A. P. Sinitsyn (2004). Specific xyloglucanases as a new class of polysaccharide-degrading enzymes. *Biochimica et Biophysica Acta (BBA)—General Subjects* **1674**(3): 268–281.

Guarnetti, R. L. (2007). *Estudo da sustentibilidade ambiental do cultivo commercial do bamboo gicante: produção de colmos e brotos.* Unpublished master's dissertation, Universidade Paulista-UNIP.

Gübitz, G. M., M. Hayn, M. Sommerauer and W. Steiner (1996). Mannan-degrading enzymes from *Sclerotium rolfsii*: Characterisation and synergism of two endo β-mannanases and a β-mannosidase. *Bioresour Technol* **58**(2): 127–135.

Gubler, F., A. E. Ashford, A. Bacic, A. B. Blakeney and B. A. Stone (1985). Release of ferulic acid from barley aleurone. II. Characterization of the feruloyl compounds released in response to GA3. *Aust J Plant Physiol* **12**: 307–317.

Guenther, C. M., B. E. Kuypers, M. T. Lam, T. M. Robinson, J. Zhao and J. Suh (2014). Synthetic virology: Engineering viruses for gene delivery. *Wiley Interdiscip Rev Nanomed Nanobiotechnol* **6**(6): 548–558.

Guillén, F., A. T. Martinez and M. J. Martínez (1990). Production of hydrogen peroxide by aryl-alcohol oxidase from the ligninolytic fungus *Pleurotus eryngii*. *Appl Microbiol Biotechnol Bioeng* **32**(4): 465–469.

Guillen, F., M. J. Martinez, C. Munoz and A. T. Martinez (1997). Quinone redox cycling in the ligninolytic fungus *Pleurotus eryngii* leading to extracellular production of superoxide anion radical. *Arch Biochem Biophys* **339**(1): 190–199.

Guillon, F. and J.-F. Thibault (1989). Methylation analysis and mild acid hydrolysis of the "hairy" fragments of sugar-beet pectins. *Carbohydr Res* **190**(1): 85–96.

Guillon, F., J.-F. Thibault, F. M. Rombouts, A. G. J. Voragen and W. Pilnik (1989). Enzymic hydrolysis of the "hairy" fragments of sugar-beet pectins. *Carbohydr Res* **190**(1): 97–108.

Guo, H. H., J. Choe and L. A. Loeb (2004). Protein tolerance to random amino acid change. *Proc Natl Acad Sci USA* **101**(25): 9205–9210.

Gupta, N. and E. T. Farinas (2010). Directed evolution of CotA laccase for increased substrate specificity using *Bacillus subtilis* spores. *Protein Eng Des Sel* **23**(8): 679–682.

Gupta, R. D. and D. S. Tawfik (2008). Directed enzyme evolution via small and effective neutral drift libraries. *Nat Methods* **5**(11): 939–942.

Gusakov, A. V. and A. P. Sinitsyn (1992). A theoretical analysis of cellulase product inhibition: Effect of cellulase binding constant, enzyme/substrate ratio, and beta-glucosidase activity on the inhibition pattern. *Biotechnol Bioeng* **40**(6): 663–671.

Gutierrez, A., L. Caramelo, A. Prieto, M. J. Martinez and A. T. Martinez (1994). Anisaldehyde production and aryl-alcohol oxidase and dehydrogenase activities in ligninolytic fungi of the genus *Pleurotus*. *Appl Environ Microbiol* **60**(6): 1783–1788.

Habibi, Y., L. A. Lucia and O. J. Rojas (2010). Cellulose nanocrystals: Chemistry, self-assembly, and applications. *Chem Rev* **110**(6): 3479–3500.

Haitjema, C., K. Solomon and M. A. O'Malley (2013). Biochemical insight into fungal cellulosome architecture and regulation. *AIChE Annual Meeting*, San Francisco, CA.

Hakkinen, M., M. Arvas, M. Oja, N. Aro, M. Penttila, M. Saloheimo and T. Pakula (2012). Re-annotation of the CAZy genes of *Trichoderma reesei* and transcription in the presence of lignocellulosic substrates. *Microb Cell Fact* **11**: 134.

Hall, J., G. P. Hazlewood, P. J. Barker and H. J. Gilbert (1988). Conserved reiterated domains in *Clostridium thermocellum* endoglucanases are not essential for catalytic activity. *Gene* **69**(1): 29–38.

Hall, M., P. Bansal, J. H. Lee, M. J. Realff and A. S. Bommarius (2010). Cellulose crystallinity—A key predictor of the enzymatic hydrolysis rate. *FEBS J* **277**(6): 1571–1582.

Hallac, B. B., P. Sannigrahi, Y. Pu, M. Ray, R. J. Murphy and A. J. Ragauskas (2010). Effect of ethanol organosolv pretreatment on enzymatic hydrolysis of *Buddleja davidii* stem biomass. *Ind Eng Chem Res* **49**(4): 1467–1472.

Halstead, J. R., P. E. Vercoe, H. J. Gilbert, K. Davidson and G. P. Hazlewood (1999). A family 26 mannanase produced by *Clostridium thermocellum* as a component of the cellulosome contains a domain which is conserved in mannanases from anaerobic fungi. *Microbiology* **145**(11): 3101–3108.

Hammel, K. E., A. N. Kapich, K. A. Jensen Jr and Z. C. Ryan (2002). Reactive oxygen species as agents of wood decay by fungi. *Enzyme Microb Technol* **30**(4): 445–453.

Hammel, K. E., M. Tien, B. Kalyanaraman and T. K. Kirk (1985). Mechanism of oxidative C alpha-C beta cleavage of a lignin model dimer by *Phanerochaete chrysosporium* ligninase. Stoichiometry and involvement of free radicals. *J Biol Chem* **260**(14): 8348–8353.

Hammel, M., H. P. Fierobe, M. Czjzek, S. Finet and V. Receveur-Brechot (2004). Structural insights into the mechanism of formation of cellulosomes probed by small angle X-ray scattering. *J Biol Chem* **279**(53): 55985–55994.

Hammel, M., H. P. Fierobe, M. Czjzek, V. Kurkal, J. C. Smith, E. A. Bayer, S. Finet and V. Receveur-Brechot (2005). Structural basis of cellulosome efficiency explored by small angle X-ray scattering. *J Biol Chem* **280**(46): 38562–38568.

Handelsman, J., M. R. Rondon, S. F. Brady, J. Clardy and R. M. Goodman (1998). Molecular biological access to the chemistry of unknown soil microbes: A new frontier for natural products. *Chem Biol* **5**(10): R245–249.

Hannuksela, T. and C. H. du Penhoat (2004). NMR structural determination of dissolved O-acetylated galactoglucomannan isolated from spruce thermomechanical pulp. *Carbohydr Res* **339**(2): 301–312.

Hardiman, E., M. Gibbs, R. Reeves and P. Bergquist (2010). Directed evolution of a thermophilic beta-glucosidase for cellulosic bioethanol production. *Appl Biochem Biotechnol* **161**(1–8): 301–312.

Harkin, J. M. and J. R. Obst (1974). Demethylation of 2,4,6-trimethoxyphenol by phenol oxidases: A model for chromophore formation in wood and pulp. *Tappi J* **57**(7): 118–121

Harris, P. V., D. Welner, K. C. McFarland, E. Re, J.-C. Navarro Poulsen, K. Brown, R. Salbo et al. (2010). Stimulation of lignocellulosic biomass hydrolysis by proteins of glycoside hydrolase family 61: Structure and function of a large, enigmatic family. *Biochemistry* **49**(15): 3305–3316.

Hartley, R. D., W. H. Morrison Iii, F. Balza and G. H. N. Towers (1990). Substituted truxillic and truxinic acids in cell walls of *Cynodon dactylon. Phytochemistry* **29**(12): 3699–3703.

Harvey, P. J., H. E. Schoemaker, R. M. Bowen and J. M. Palmer (1985). Single-electron transfer processes and the reaction mechanism of enzymic degradation of lignin. *FEBS Lett* **183**(1): 13–16.

Hatakeyama, H. and T. Hatakeyama (2010). Lignin structure, properties, and applications. In *Biopolymers. Lignin, Proteins, Bioactive Nanocomposites*, A. Abe, K. Dusek and S. Kobayashi (Eds.), Vol. 232, pp. 1–63. Berlin, Germany: Springer-Verlag.

Hatakeyama, T. and H. Hatakeyama (1982). Temperature dependence of X-ray diffractograms of amorphous lignins and polystyrenes *Polymer* **23**(3): 475–477.

Hatakka, A. and K. E. Hammel (2010). Fungal biodegradation of lignocelluloses. In *The Mycota, A Comprehensive Treatise on Fungi as Experimental Systems for Basic and Applied Research*, K. Esser and J. W. Bennett (Eds.), Vol. 10, pp. 319–340. New York: Springer.

Hayashi, H., K. I. Takagi, M. Fukumura, T. Kimura, S. Karita, K. Sakka and K. Ohmiya (1997). Sequence of *xynC* and properties of XynC, a major component of the *Clostridium thermocellum* cellulosome. *J Bacteriol* **179**(13): 4246–4253.

Hayashi, H., M. Takehara, T. Hattori, T. Kimura, S. Karita, K. Sakka and K. Ohmiya (1999). Nucleotide sequences of two contiguous and highly homologous xylanase genes *xynA* and *xynB* and characterization of XynA from *Clostridium thermocellum. Appl Microbiol Biotechnol* **51**(3): 348–357.

Hazlewood, G. P., K. Davidson, J. H. Clarke, A. J. Durrant, J. Hall and H. J. Gilbert (1990). Endoglucanase E, produced at high level in *Escherichia coli* as a *lacZ'* fusion protein, is part of the *Clostridium thermocellum* cellulosome. *Enzyme Microb Technol* **12**(9): 656–662.

He, M. X., J. L. Wang, H. Qin, Z. X. Shui, Q. L. Zhu, B. Wu, F. R. Tan et al. (2014). Bamboo: A new source of carbohydrate for biorefinery. *Carbohydr Polym* **111**: 645–654.

Heaton, E. A., F. G. Dohleman and S. P. Long (2008). Meeting US biofuel goals with less land: The potential of Miscanthus. *Global Change Biol* **14**(9): 2000–2014.

Hebraud, M. and M. Fevre (1990). Purification and characterization of an extracellular beta-xylosidase from the rumen anaerobic fungus *Neocallimastix frontalis. FEMS Microbiol Lett* **60**(1–2): 11–16.

Heinzelman, P., R. Komor, A. Kanaan, P. Romero, X. Yu, S. Mohler, C. Snow and F. Arnold (2010). Efficient screening of fungal cellobiohydrolase class I enzymes for thermostabilizing sequence blocks by SCHEMA structure-guided recombination. *Protein Eng Des Sel* **23**(11): 871–880.

Heinzelman, P., C. D. Snow, M. A. Smith, X. Yu, A. Kannan, K. Boulware, A. Villalobos, S. Govindarajan, J. Minshull and F. H. Arnold (2009a). SCHEMA recombination of a fungal cellulase uncovers a single mutation that contributes markedly to stability. *J Biol Chem* **284**(39): 26229–26233.

Heinzelman, P., C. D. Snow, I. Wu, C. Nguyen, A. Villalobos, S. Govindarajan, J. Minshull and F. H. Arnold (2009b). A family of thermostable fungal cellulases created by structure-guided recombination. *Proc Natl Acad Sci USA* **106**(14): 5610–5615.

Henriksson, G., G. Johansson and G. Pettersson (2000). A critical review of cellobiose dehydrogenases. *J Biotechnol* **78**(2): 93–113.

Henriksson, G., A. Nutt, H. Henriksson, B. Pettersson, J. Stahlberg, G. Johansson and G. Pettersson (1999). Endoglucanase 28 (Cel12A), a new *Phanerochaete chrysosporium* cellulase. *Eur J Biochem* **259**(1–2): 88–95.

Henriksson, G., G. Pettersson, G. Johansson, A. Ruiz and E. Uzcategui (1991). Cellobiose oxidase from *Phanerochaete chrysosporium* can be cleaved by papain into two domains. *Eur J Biochem* **196**(1): 101–106.

Henrissat, B., H. Driguez, C. Viet and M. Schulein (1985). Synergism of cellulases from *Trichoderma reesei* in the degradation of cellulose. *Nat Biotech* **3**(8): 722–726.

Hernandez, K. and R. Fernandez-Lafuente (2011). Control of protein immobilization: Coupling immobilization and site-directed mutagenesis to improve biocatalyst or biosensor performance. *Enzyme Microb Technol* **48**(2): 107–122.

Hernandez-Ortega, A., P. Ferreira and A. T. Martinez (2012). Fungal aryl-alcohol oxidase: A peroxide-producing flavoenzyme involved in lignin degradation. *Appl Microbiol Biotechnol* **93**(4): 1395–1410.

Herpoel-Gimbert, I., A. Margeot, A. Dolla, G. Jan, D. Molle, S. Lignon, H. Mathis, J.-C. Sigoillot, F. Monot and M. Asther (2008). Comparative secretome analyses of two *Trichoderma reesei* RUT-C30 and CL847 hypersecretory strains. *Biotechnol Biofuels* **1**(1): 18.

Herrmann, M. C., M. Vrsanska, M. Jurickova, J. Hirsch, P. Biely and C. P. Kubicek (1997). The beta-D-xylosidase of *Trichoderma reesei* is a multifunctional beta-D-xylan xylohydrolase. *Biochem J* **321**(Pt 2): 375–381.

Hervé, C., A. Rogowski, A. W. Blake, S. E. Marcus, H. J. Gilbert and J. P. Knox (2010). Carbohydrate-binding modules promote the enzymatic deconstruction of intact plant cell walls by targeting and proximity effects. *Proc Natl Acad Sci USA* **107**(34): 15293–15298.

Hill, J., E. Nelson, D. Tilman, S. Polasky and D. Tiffany (2006). Environmental, economic, and energetic costs and benefits of biodiesel and ethanol biofuels. *Proceedings of the National Academy of Sciences* **103**(30): 11206–11210.

Himmel, M. E., S. Y. Ding, D. K. Johnson, W. S. Adney, M. R. Nimlos, J. W. Brady and T. D. Foust (2007). Biomass recalcitrance: Engineering plants and enzymes for biofuels production. *Science* **315**(5813): 804–807.

Hjortkjaer, R. K., V. Bille-Hansen, K. P. Hazelden, M. McConville, D. B. McGregor, J. A. Cuthbert, R. J. Greenough, E. Chapman, J. R. Gardner and R. Ashby (1986). Safety evaluation of celluclast®, an acid cellulase derived from *Trichoderma reesei*. *Food Chem Toxicol* **24**(1): 55–63.

Hodkinson, T. R., M. W. Chase, C. Takahashi, I. J. Leitch, M. D. Bennett and S. A. Renvoize (2002). The use of *dna* sequencing (ITS and *trnL-F*), AFLP, and fluorescent in situ hybridization to study allopolyploid *Miscanthus* (Poaceae). *Am J Bot* **89**(2): 279–286.

Hoffman, M., Z. Jia, M. J. Peña, M. Cash, A. Harper, A. R. N. Blackburn, A. Darvill and W. S. York (2005). Structural analysis of xyloglucans in the primary cell walls of plants in the subclass *Asteridae*. *Carbohydr Res* **340**(11): 1826–1840.

Hokanson, C. A., G. Cappuccilli, T. Odineca, M. Bozic, C. A. Behnke, M. Mendez, W. J. Coleman and R. Crea (2011). Engineering highly thermostable xylanase variants using an enhanced combinatorial library method. *Protein Eng Des Sel* **24**(8): 597–605.

Holtzapple, M., M. Cognata, Y. Shu and C. Hendrickson (1990). Inhibition of *Trichoderma reesei* cellulase by sugars and solvents. *Biotechnol Bioeng* **36**(3): 275–287.

Holtzapple, M. T., H. S. Caram and A. E. Humphrey (1984). Determining the inhibition constants in the HCH-1 model of cellulose hydrolysis. *Biotechnol Bioeng* **26**(7): 753–757.

Hon, D. N.-S. (1994). Cellulose: A random walk along its historical path. *Cellulose* **1**: 1–25.

Honda, Y. and M. Kitaoka (2004). A family 8 glycoside hydrolase from *Bacillus halodurans* C-125 (BH2105) is a reducing end xylose-releasing exo-oligoxylanase. *J Biol Chem* **279**(53): 55097–550103.

Hong, J., X. Ye and Y. H. Zhang (2007). Quantitative determination of cellulose accessibility to cellulase based on adsorption of a nonhydrolytic fusion protein containing CBM and GFP with its applications. *Langmuir* **23**(25): 12535–12540.

Hong, M. R., C. S. Park and D. K. Oh (2009). Characterization of a thermostable endo-1,5-alpha-L-arabinanase from *Caldicellulorsiruptor saccharolyticus*. *Biotechnol Lett* **31**(9): 1439–1443.

Hongshu, Z., Y. Jinggan and Z. Yan (2002). The glucomannan from ramie. *Carbohydr Polym* **47**(1): 83–86.

Hongzhang, C. and L. Liying (2007). Unpolluted fractionation of wheat straw by steam explosion and ethanol extraction. *Bioresour Technol* **98**(3): 666–676.

Horton, G. L., D. B. Rivers and G. H. Emert (1980). Preparation of cellulosics for enzymatic conversion. *Ind Eng Chem Prod Res Dev* **19**: 422–429.

Hossain, M. M. and L. Aldous (2012). Ionic liquids for lignin processing: Dissolution, isolation, and conversion. *Aust J Chem* **65**: 1465–1477.

Howard, J. L. (2012). U.S. timber production, trade, consumption, and price statistics, 1965-2010. Table 8b. Madison, WI: U.S., Department of Agriculture, Forest Service, Forest Products Laboratory.

Hsieh, C. W., D. Cannella, H. Jørgensen, C. Felby and L. G. Thygesen (2014). Cellulase inhibition by high concentrations of monosaccharides. *J Agric Food Chem* **62**(17): 3800–3005.

Hu, J., V. Arantes and J. N. Saddler (2011). The enhancement of enzymatic hydrolysis of lignocellulosic substrates by the addition of accessory enzymes such as xylanase: Is it an additive or synergistic effect? *Biotechnol Biofuels* **4**: 36.

Hu, J., R. P. Chandra, V. Arantes, K. Gourlay, J. S. V. Dyk and J. N. Saddler (2015). The addition of accessory enzymes enhances the hydrolytic performance of cellulase enzymes at high solid loadings. *Bioresour Technol* **186**: 149–153.

Hu, Z. and Z. Wen (2008). Enhancing enzymatic digestibility of switchgrass by microwave-assisted alkali pretreatment. *Biochem Eng J* **38**(3): 369–378.

Huang, J.-W., C.-C. Chen, C.-H. Huang, T.-Y. Huang, T.-H. Wu, Y.-S. Cheng, T.-P. Ko, C.-Y. Lin, J.-R. Liu and R.-T. Guo (2014). Improving the specific activity of β-mannanase from *Aspergillus niger* BK01 by structure-based rational design. *Biochim Biophys Acta Proteins Proteomics* **1844**(3): 663–669.

Huberman, M. A. (1959). Bamboo silviculture. In *An International Review of Forestry and Forest Products*, Unasylva. Forestry and Forest Products Division. Food & Agriculture Organization of the United Nations, p. 13.

Huddleston, J. G., A. E. Visser, W. M. Reichert, H. D. Willauer, G. A. Broker and R. D. Rogers (2001). Characterization and comparison of hydrophilic and hydrophobic room temperature ionic liquids incorporating the imidazolium cation. *Green Chem* 3(4): 156–164.

Ibrahim, V., L. Mendozaa, G. Mamo and R. Hatti-Kaul (2011). Blue laccase from *Galerina* sp.: Properties and potential for Kraft lignin demethylation. *Process Biochem* 46(1): 379–384.

Igarashi, K., A. Koivula, M. Wada, S. Kimura, M. Penttilä and M. Samejima (2009). High speed atomic force microscopy visualizes processive movement of *Trichoderma reesei* cellobiohydrolase I on crystalline cellulose. *J Biol Chem* 284(52): 36186–36190.

Igarashi, K., M. Samejima and K. E. L. Eriksson (1998). Cellobiose dehydrogenase enhances *Phanerochaete chrysosporium* cellobiohydrolase I activity by relieving product inhibition. *Eur J Biochem* 253(1): 101–106.

Igarashi, K., T. Uchihashi, A. Koivula, M. Wada, S. Kimura, T. Okamoto, M. Penttilä, T. Ando and M. Samejima (2011). Traffic jams reduce hydrolytic efficiency of cellulase on cellulose surface. *Science* 333(6047): 1279–1282.

Igual, J. M., E. Velázquez, P. F. Mateos, C. Rodríguez-Barrueco, E. Cervantes and E. Martínez-Molina (2001). Cellulase isoenzyme profiles in Frankia strains belonging to different cross-inoculation groups. *Plant Soil* 229(1): 35–39.

Iiyama, K., T. Lam and B. A. Stone (1994). Covalent cross-links in the cell wall. *Plant Physiol* 104(2): 315–320.

Inglett, G. E. (1970). Kernel structure, composition, and quality. In *Corn: Culture, Processing, Products*, G. E. Inglett (Ed.). Westport, CA: Avi Publishing Co.

International Energy Agency (2011). Technology roadmap—biofuels for transport. Retrieved from https://www.iea.org/publications/freepublications/publication/name,3976,en.html.

International Renewable Energy Agency (2013). *Renewable Energy and Jobs*. Abu Dhabi, United Arab Emirates: IRENA.

Irwin, D., D.-H. Shin, S. Zhang, B. K. Barr, J. Sakon, P. A. Karplus and D. B. Wilson (1998). Roles of the catalytic domain and two cellulose binding domains of *Thermomonospora fusca* E4 in cellulose hydrolysis. *J Bacteriol* 180(7): 1709–1714.

Ishihara, T. and M. Miyaxaki (1974). Demethylation of lignin and lignin models by fungal laccase. *Mokurai Gakkaishi* 20(1): 39.

Ishihara, T. and M. Miyaxaki (1976). Oxidation of syringic acid beta oxidation of syringic acid by fungal laccase y fungal laccase. *Mokuzai Gakkoishi* 22: 371.

Ishii, T. and T. Hiroi (1990). Linkage of phenolic acids to cell-wall polysaccharides of bamboo shoot. *Carbohydr Res* 206(2): 297–310.

Ishii, T., T. Matsunaga, P. Pellerin, M. A. O'Neill, A. Darvill and P. Albersheim (1999). The plant cell wall polysaccharide rhamnogalacturonan II self-assembles into a covalently cross-linked dimer. *J Biol Chem* 274(19): 13098–13104.

Isholaa, M. M., Isroic and M. J. Taherzadeha (2014). Effect of fungal and phosphoric acid pretreatment on ethanol production from oil palm empty fruit bunches (OPEFB). *Bioresour Technol* 165: 9–12.

Islam, S. M. M., J. R. Elliott and L.-K. Ju (2014). Supercritical CO_2 based pretreatment of guayule biomass for high sugar yield and low inhibition hydrolysate. *2014 AIChE Annual Meeting*, Atlanta, GA.

Itoh, H., M. Wada, Y. Honda, M. Kuwahara and T. Watanabe (2003). Bioorganosolve pretreatments for simultaneous saccharification and fermentation of beech wood by ethanolysis and white rot fungi. *J Biotechnol* **103**(3): 273–280.

Iwai, H., T. Ishii and S. Satoh (2001). Absence of arabinan in the side chains of the pectic polysaccharides strongly associated with cell walls of *Nicotiana plumbaginifolla* non-organogenic callus with loosely attached constituent cells. *Planta* **213**(6): 907–915.

Jackowiak, D., D. Bassard, A. Pauss and T. Ribeiro (2011). Optimisation of a microwave pretreatment of wheat straw for methane production. *Bioresour Technol* **102**(12): 6750–6756.

Jamal-Talabani, S., A. B. Boraston, J. P. Turkenburg, N. Tarbouriech, V. M. A. Ducros and G. J. Davies (2004). Ab Initio structure determination and functional characterization of CBM36: A new family of calcium-dependent carbohydrate binding modules. *Structure* **12**(7): 1177–1187.

Janssen, F. W., R. M. Kerwin and H. W. Ruelius (1965). Alcohol oxidase, a novel enzyme from a basidiomycete. *Biochem Biophys Res Commun* **20**(5): 630–634.

Janssen, F. W. and H. W. Ruelius (1968a). Carbohydrate oxidase, a novel enzyme from polyporus obtusus: II. Specificity and characterization of reaction products. *Biochim Biophys Acta* **167**(3): 501–510.

Janssen, F. W. and H. W. Ruelius (1968b). Alcohol oxidase, a flavoprotein from several Basidiomycetes species. Crystallization by fractional precipitation with polyethylene glycol. *Biochim Biophys Acta* **151**: 303–342.

Jarvis, M. C. (2011). Plant cell walls: Supramolecular assemblies. *Food Hydrocoll* **25**(2): 257–262.

Jensen, K. A., W. Bao, S. Kawai, E. Srebotnik and K. E. Hammel (1996). Manganese-dependent cleavage of nonphenolic lignin structures by *Ceriporiopsis subvermispora* in the absence of lignin peroxidase. *Appl Environ Microbiol* **62**(10): 3679–3686.

Jeoh, T. (1998). *Steam explosion pretreatment of cotton gin waste for fuel ethanol production*. Master of science Master's thesis, Virginia Tech. University.

Ji, L., J. Yang, H. Fan, Y. Yang, B. Li, X. Yu, N. Zhu and H. Yuan (2014). Synergy of crude enzyme cocktail from cold-adapted *Cladosporium cladosporioides* Ch2-2 with commercial xylanase achieving high sugars yield at low cost. *Biotechnol Biofuels* **7**: 130.

Jimenéz-Zurdo, J. I., P. F. Mateos, F. B. Dazzo and E. Martínez-Molina (1996). Cell-bound cellulase and polygaracturonase production by *Rhizobium* and *Bradyrhizobium* species. *Soil Biol Biochem* **28**(7): 917–921.

Jindou, S., A. Soda, S. Karita, T. Kajino, P. Béguin, J. H. D. Wu, M. Inagaki, T. Kimura, K. Sakka and K. Ohmiya (2004). Cohesin-dockerin interactions within and between *Clostridium josui* and *Clostridium thermocellum*: Binding seectivity between cognate dockerin and cohesin domains and species specificity. *J Biol Chem* **279**(11): 9867–9874.

Jochens, H. and U. T. Bornscheuer (2010). Natural diversity to guide focused directed evolution. *Chembiochem* **11**(13): 1861–1866.

Johansson, T., K. G. Welinder and P. O. Nyman (1993). Isozymes of lignin peroxidase and manganese(II) peroxidase from the white-rot Basidiomycete *Trametes versicolor*: II. Partial sequences, peptide maps, and amino acid and carbohydrate compositions. *Arch Biochem Biophys* **300**(1): 57–62.

Johnson, D. L., J. L. Thompson, S. M. Brinkmann, K. A. Schuller and L. L. Martin (2003). Electrochemical characterization of purified *Rhus vernicifera* Laccase: Voltammetric evidence for a sequential four-electron transfer. *Biochemistry* **42**: 10229–10237.

Johnson, E. A., M. Sakajoh, G. Halliwell, A. Madia and A. L. Demain (1982). Saccharification of complex cellulosic substrates by the cellulase system from *Clostridium thermocellum*. *Appl Environ Microbiol* **43**(5): 1125–1132.

Johnson, R. L. (1972). Genetically improved cottonwood: A research and development success. *Proceedings of the 1972 National Convention, Society of Americans Foresters*, pp. 113–119. Washington, DC: Society of American Foresters.

Johnsson, K., R. K. Allemann, H. Widmer and S. A. Benner (1993). Synthesis, structure and activity of artificial, rationally designed catalytic polypeptides. *Nature* **365**(6446): 530–532.

Joliff, G., P. Béguin and J.-P. Aubert (1986a). Nucleotide sequence of the cellulase gene *celD* encoding endoglucanase D of *Clostridium thermocellum*. *Nucleic Acids Res* **14**(21): 8605–8612.

Joliff, G., P. Beguin, M. Juy, J. Millet, A. Ryter, R. Poljak and J.-P. Aubert (1986b). Isolation, crystallization and properties of a new cellulase of *Clostridium thermocellum* overproduced in *Escherichia coli*. *Nat Biotech* **4**(10): 896–900.

Joo, J. C., S. P. Pack, Y. H. Kim and Y. J. Yoo (2011). Thermostabilization of *Bacillus circulans* xylanase: Computational optimization of unstable residues based on thermal fluctuation analysis. *J Biotechnol* **151**(1): 56–65.

Joo, J. C., S. Pohkrel, S. P. Pack and Y. J. Yoo (2010). Thermostabilization of *Bacillus circulans* xylanase via computational design of a flexible surface cavity. *J Biotechnol* **146**(1–2): 31–39.

Jørgensen, H., J. B. Kristensen and C. Felby (2007). Enzymatic conversion of lignocellulose into fermentable sugars: Challenges and opportunities. *Biofuels Bioprod Biorefin* **1**(2): 119–134.

Jorgenson, L. (1950). *Studies on the Partial Hydrolysis of Cellulose*. Oslo, Norway: Moestue.

Joshi, S., O. Sakamoto and H. L. MacLean (2005). Economic feasibility analysis of municipal solid waste to ethanol conversion. *27th Symposium on Biotechnology for Fuels and Chemicals*, Denver, CO.

Juge, N., M. L. Gal-Coeffet, C. Furniss, A. Gunning, B. Kramhoft, V. J. Morris, G. Williamson and B. Svensson (2002). The starch binding domain of glucoamylase from *Aspergillus niger*: Overview of its structure, function, and role in raw-starch hydrolysis. *Biologia (Bratisl)* **57**: 239–245.

Juhasz, T., Z. Szengyel, K. Reczey, M. Siika-Aho and L. Viikari (2005). Characterization of cellulases and hemicellulases produced by *Trichoderma reesei* on various carbon sources. *Process Biochem* **40**: 3519–3525.

Jun, H., H. Guangye and C. Daiwen (2013). Insights into enzyme secretion by filamentous fungi: Comparative proteome analysis of *Trichoderma reesei* grown on different carbon sources. *J Proteomics*. **89**: 191–201.

Jung, K. H., J.-H. Lee, Y.-T. Yi, H.-K. Kim and M.-Y. Pack (1992). Properties of a novel *Clostridium thermocellum* endo-β-1,4-glucanase expressed in *Escherichia coli*. *Kor J Appl Microbiol Biotechnol* **20**: 505–510.

Jung, Y. H., I. J. Kim, H. K. Kim and K. H. Kim (2013). Dilute acid pretreatment of lignocellulose for whole slurry ethanol fermentation. *Bioresour Technol* **132**: 109–114.

Jürgens, C., A. Strom, D. Wegener, S. Hettwer, M. Wilmanns and R. Sterner (2000). Directed evolution of a (βα)8-barrel enzyme to catalyze related reactions in two different metabolic pathways. *Proc Natl Acad Sci USA* **97**(18): 9925–9930.

Juturu, V., T. M. Teh and J. C. Wu (2014). Expression of Aeromonas punctata ME-1 exo-xylanase X in *E. coli* for efficient hydrolysis of xylan to xylose. *Appl Biochem Biotechnol* **174**(8): 2653–2662.

Kabo, G. J., A. V. Blokhin, Y. U. Paulechka, A. G. Kabo, M. P. Shymanovich and J. W. Magee (2004). Thermodynamic properties of 1-butyl-3-methylimidazolium hexafluorophosphate in the condensed state. *J Chem Eng Data* **49**(3): 453–461.

Kakiuchi, M., A. Isui, K. Suzuki, T. Fujino, E. Fujino, T. Kimura, S. Karita, K. Sakka and K. Ohmiya (1998). Cloning and DNA sequencing of the genes encoding *Clostridium josui* scaffolding protein CipA and cellulase CelD and identification of their gene products as major components of the cellulosome. *J Bacteriol* **180**(16): 4303–4308.

Kamat, R. K., W. Ma, Y. Yang, Y. Zhang, C. Wang, C. V. Kumar and Y. Lin (2013). Adsorption and hydrolytic activity of the polycatalytic cellulase nanocomplex on cellulose. *ACS Appl Mater Interfaces* **5**(17): 8486–8494.

Kang, K. E., G. T. Jeong, C. Sunwoo and D. H. Park (2012). Pretreatment of rapeseed straw by soaking in aqueous ammonia. *Bioprocess Biosyst Eng* **35**: 77–84.

Kapich, A. N., K. T. Steffen, M. Hofrichter and A. Hatakka (2005). Involvement of lipid peroxidation in the degradation of a non-phenolic lignin model compound by manganese peroxidase of the litter-decomposing fungus Stropharia coronilla. *Biochem Biophys Res Commun* **330**(2): 371–377.

Kapich, A. N., T. V. Korneichik, A. Hatakka and K. E. Hammel (2010). Oxidizability of unsaturated fatty acids and of a non-phenolic lignin structure in the manganese peroxidase-dependent lipid peroxidation system. *Enzyme Microb Technol* **46**: 136–140.

Karimia, K., S. Kheradmandinia and M. J. Taherzadeh (2006). Conversion of rice straw to sugars by dilute-acid hydrolysis. *Biomass Bioenergy* **30**: 247–253.

Karlsson, J., M. Siika-aho, M. Tenkanen and F. Tjerneld (2002). Enzymatic properties of the low molecular mass endoglucanases Cel12A (EG III) and Cel45A (EG V) of *Trichoderma reesei*. *J Biotechnol* **99**: 63–78.

Karpol, A., Y. Barak, R. Lamed, Y. Shoham and E. A. Bayer (2008). Functional asymmetry in cohesin binding belies inherent symmetry of the dockerin module: Insight into cellulosome assembly revealed by systematic mutagenesis. *Biochem J* **410**(2): 331–338.

Kataeva, I., X.-L. Li, H. Chen, S.-K. Choi and L. G. Ljungdahl (1999). Cloning and sequence analysis of a new cellulase gene encoding CelK, a major cellulosome component of *Clostridium thermocellum*: Evidence for gene duplication and recombination. *J Bacteriol* **181**(17): 5288–5295.

Kataria, R. and S. Ghosh (2011). Saccharification of Kans grass using enzyme mixture from *Trichoderma reesei* for bioethanol production. *Bioresour Technol* **102**(21): 9970–9975.

Katrolia, P., Q. Yan, P. Zhang, P. Zhou, S. Yang and Z. Jiang (2013). Gene cloning and enzymatic characterization of an alkali-tolerant endo-1,4-β-mannanase from *Rhizomucor miehei*. *J Agric Food Chem* **61**(2): 394–401.

Kauffman, S. A. (1993). *The Origins of Order: Self Organization and Selection in Evolution*. New York: Oxford University Press.

Kawai, S., T. Umezawa and T. Higuchi (1988). Degradation mechanisms of phenolic β-1 lignin substructure model compounds by laccase of *Coriolus versicolor*. *Arch Biochem Biophys* **262**(1): 99–110.

Kazlauskas, R. J. and U. T. Bornscheuer (2009). Finding better protein engineering strategies. *Nat Chem Biol* **5**(8): 526–569.

Keegstra, K., K. W. Talmadge, W. D. Bauer and P. Albersheim (1973). The structure of plant cell walls III. A model of the walls of suspension-cultured sycamore cells based on the interconnections of the macromolecular components. *Plant Physiol* **51**(1): 188–197.

Kelley, R. L. and C. A. Reddy (1986). Identification of glucose oxidase activity as the primary source of hydrogen peroxide production in ligninolytic cultures of *Phanerochaete chrysosporium. Arch Microbiol* **144**(3): 248–253.

Kempton, J. B. and S. G. Withers (1992). Mechanism of *Agrobacterium* β-glucosidase: Kinetic studies. *Biochemistry* **31**: 9961–9969.

Kersten, P. J. (1990). Glyoxal oxidase of *Phanerochaete chrysosporium*: Its characterization and activation by lignin peroxidase. *Proc Natl Acad Sci USA* **87**(8): 2936–2940.

Kersten, P. J. and T. K. Kirk (1987). Involvement of a new enzyme, glyoxal oxidase, in extracellular H_2O_2 production by *Phanerochaete chrysosporium. J Bacteriol* **169**(5): 2195–2201.

Kester, H. C. M., J. A. E. Benen and J. Visser (1999a). The exopolygalacturonase from *Aspergillus tubingensis* is also active on xylogalacturonan. *Biotechnol Appl Biochem* **30**: 53–57.

Kester, H. C. M., D. Magaud, C. Roy, D. Anker, A. Doutheau, V. Shevchik, N. Hugouvieux-Cotte-Pattat, J. A. Benen and J. Visser (1999b). Performance of selected microbial pectinases on synthetic monomethyl-esterified di- and trigalacturonates. *J Biol Chem* **274**(52): 37053–37059.

Khalili, B., F. Nourbakhsh, N. Nili, H. Khademi and B. Sharifna (2011). Diversity of soil cellulase isoenzymes is associated with soil cellulase kinetic and thermodynamic parameters. *Soil Biol Biochem* **43**(8): 1639–1648.

Khan, A. W., J. Labrie and J. McKeown (1987). Electron-beam irradation pretreatment and enzymatic saccharification of used newsprint and paper-mill wastes. *Rad Phys Chem* **29**(2): 117–120.

Kim, I., Y. H. Seo, G.-Y. Kim and J.-I. Han (2015a). Co-production of bioethanol and biodiesel from corn stover pretreated with nitric acid. *Fuel* **143**: 285–289.

Kim, J., S. Kim, S. Yoon, E. Hong and Y. Ryu (2015b). Improved enantioselectivity of thermostable esterase from *Archaeoglobus fulgidus* toward (S)-ketoprofen ethyl ester by directed evolution and characterization of mutant esterases. *Appl Microbiol Biotechnol* **99**(15): 6293–6301.

Kim, K. H. and J. Hong (2001). Supercritical CO_2 pretreatment of lignocellulose enhances enzymatic cellulose hydrolysis. *Bioresour Technol* **77**(2): 139–144.

Kim, S. and B. E. Dale (2004). Global potential bioethanol production from wasted crops and crop residues. *Biomass Bioenergy* **26**(4): 361–375.

Kim, S., J. M. Park, J.-W. Seo and C. H. Kim (2012). Sequential acid-/alkali-pretreatment of empty palm fruit bunch fiber. *Bioresour Technol* **109**: 229–233.

Kim, S. J., K. Ishikawa, M. Hirai and M. Shoda (1995). Characteristics of a newly isolated fungus, *Geotrichum candidum* Dec 1, which decolorizes various dyes. *J Ferment Bioeng* **79**(6): 601–607.

Kim, S. J. and M. Shoda (1999). Purification and characterization of a novel peroxidase from *Geotrichum candidum* Dec 1 involved in decolorization of dyes. *Appl Environ Microbiol* **65**(3): 1029–1035.

Kim, T. H., J. S. Kim, C. Sunwoo and Y. Y. Lee (2003). Pretreatment of corn stover by aqueous ammonia. *Bioresour Technol* **90**(1): 39–47.

Kim, T. H. and Y. Y. Lee (2005a). Pretreatment of corn stover by soaking in aqueous ammonia. *Appl Biochem Biotechnol* **121–124**: 1119–1131.

Kim, T. H. and Y. Y. Lee (2005b). Pretreatment and fractionation of corn stover by ammonia recycle percolation process. *Bioresour Technol* **96**(18): 2007–2013.

Kim, T. H. and Y. Y. Lee (2006). Fractionation of corn stover by hot-water and aqueous ammonia treatment. *Bioresour Technol* **97**(2): 224–232.

Kim, Y., R. Hendrickson, N. S. Mosier and M. R. Ladisch (2009a). Liquid hot water pretreatment of cellulosic biomass. *Methods Mol Biol* **581**: 93–102.

Kim, Y., N. S. Mosier and M. R. Ladisch (2009b). Enzymatic digestion of liquid hot water pretreated hybrid poplar. *Biotechnol Prog* **25**(2): 340–348.

Kim, Y.-S., H.-C. Jung and J.-G. Pan (2000). Bacterial cell surface display of an enzyme library for selective screening of improved cellulase variants. *Appl Environ Microbiol* **66**(2): 788–793.

Kirby, R. (2006). Actinomycetes and lignin degradation. *Adv Appl Microbiol* **58**: 125–168.

Kirikyali, N. and I. F. Connerton (2014). Heterologous expression and kinetic characterisation of *Neurospora crassa* β-xylosidase in *Pichia pastoris*. *Enzyme Microb Technol* **57**: 63–68.

Klein-Marcuschamer, D., P. Oleskowicz-Popiel, B. Simmons and H. Blanch (2012). The challenge of enzyme cost in the production of lignocellulosic biofuels. *Biotechnol Bioeng* **109**(4): 1083–1087.

Kleman-Leyer, K. M., M. Siika-Aho, T. T. Teeri and T. K. Kirk (1996). The cellulases endoglucanase I and cellobiohydrolase II of *Trichoderma reesei* act synergistically to solubilize native cotton cellulose but not to decrease its molecular size. *Appl Environ Microbiol* **62**(8): 2883–2887.

Knowles, J., P. Lehtovaara and T. Teeri (1987). Cellulase families and their genes. *Trends Biotechnol* **5**: 255–261.

Kobayashi, T., M. P. M. Romaniec, P. J. Barker, U. T. Gerngross and A. L. Demain (1993). Nucleotide sequence of gene *celM* encoding a new endoglucanase (CelM) of *Clostridium thermocellum* and purification of the enzyme. *J Ferm Bioeng* **76**(4): 251–256.

Kofod, L. V., S. Kauppinen, S. Christgau, L. N. Andersen, H. P. Heldt-Hansen, K. Dörreich and H. Dalbøge (1994). Cloning and characterization of two structurally and functionally divergent rhamnogalacturonases from *Aspergillus aculeatus*. *J Biol Chem* **269**(46): 29182–29189.

Kohlmann, K. L., P. J. Westgate, A. Sarikaya, A. Velayudhan, J. Weil, R. Hendrickson and M. R. Ladisch (1995). Enhanced enzyme activities on hydrated lignocellulosic substrates. In *Enzymatic Degradation of Insoluble Carbohydrates*, ACS Symposium Series, Vol. 618, pp. 237–255.

Kohring, S., J. Wiegel and F. Mayer (1990). Subunit composition and glycosidic activities of the cellulase complex from *Clostridium thermocellum* JW20. *Appl Environ Microbiol* **56**(12): 3798–3804.

Koivula, A., T. Kinnari, V. Harjunpaa, L. Ruohonen, A. Teleman, T. Drakenberg, J. Rouvinen, T. A. Jones and T. T. Teeri (1998). Tryptophan 272: An essential determinant of crystalline cellulose degradation by *Trichoderma reesei* cellobiohydrolase Cel6A. *FEBS Lett* **429**(3): 341–346.

Koksharov, M. I. and N. N. Ugarova (2011). Thermostabilization of firefly luciferase by in vivo directed evolution. *Protein Eng Des Sel* **24**(11): 835–844.

Komor, R. S., P. A. Romero, C. B. Xie and F. H. Arnold (2012). Highly thermostable fungal cellobiohydrolase I (Cel7A) engineered using predictive methods. *Protein Eng Des Sel* **25**(12): 827–833.

Kormelink, F. J. M., M. J. F. Searle-Van Leeuwen, T. M. Wood and A. G. J. Voragen (1991). (1,4)-β-d-Arabinoxylan arabinofuranohydrolase: A novel enzyme in the bioconversion of arabinoxylan. *Appl Microbiol Biotechnol* **35**(2): 231–232.

Koshland, D. E. (1953). Stereochemestry and the mechanism of enzymatic reactions. *Biol Rev* **28**(4): 416–436.

Kosugi, A., K. Murashima and R. H. Doi (2002). Xylanase and acetyl xylan esterase activities of XynA, a key subunit of the *Clostridium cellulovorans* cellulosome for xylan degradation. *Appl Environ Microbiol* **68**(12): 6399–6402.

Kothari, V., P. Surt and D. Kapadia (2011). Cellulosomes—A robust machinery for cellulose degradation. *IJLST* **4**(5): 31–36.

Kremer, S. M. and P. M. Wood (1992a). Production of Fenton's reagent by cellobiose oxidase from cellulolytic cultures of *Phanerochaete chrysosporium*. *Eur J Biochem* **208**(3): 807–814.

Kremer, S. M. and P. M. Wood (1992b). Evidence that cellobiose oxidase from *Phanerochaete chrysosporium* is primarily an Fe(III) reductase. Kinetic comparison with neutrophil NADPH oxidase and yeast flavocytochrome b2. *Eur J Biochem* **205**(1): 133–138.

Krishnan, M. S., N. P. Nghiem and B. H. Davison (1999). Ethanol production from corn starch in a fluidized-bed bioreactor. *Appl Biochem Biotechnol* **77– 79**(1–3): 359–372.

Kristufek, D., S. Zeilinger and C. P. Kubicek (1995). Regulation of β-xylosidase formation by xylose in *Trichoderma reesei*. *Appl Microbiol Biotechnol* **42**(5): 713–717.

Kruus, K., A. C. Lua, A. L. Demain and J. H. Wu (1995). The anchorage function of CipA (CelL), a scaffolding protein of the *Clostridium thermocellum* cellulosome. *Proc Natl Acad Sci USA* **92**(20): 9254–9258.

Kubata, B. K., T. Suzuki, H. Horitsu, K. Kawai and K. Takamizawa (1994). Purification and characterization of *Aeromonas caviae* ME-1 xylanase V, which produces exclusively xylobiose from xylan. *Appl Environ Microbiol* **60**(2): 531–535.

Kubata, B. K., K. Takamizawa, K. Kawai, T. Suzuki and H. Horitsu (1995). Xylanase IV, an exoxylanase of *Aeromonas caviae* ME-1 which produces xylotetraose as the only low-molecular-weight oligosaccharide from xylan. *Appl Environ Microbiol* **61**(4): 1666–1668.

Kuhls, K., E. Lieckfeldt, G. J. Samuels, W. Kovacs, W. Meyer, O. Petrini, W. Gams, T. Börner and C. P. Kubicek (1996). Molecular evidence that the asexual industrial fungus *Trichoderma reesei* is a clonal derivative of the ascomycete Hypocrea jecorina. *Proc Natl Acad Sci USA* **93**(15): 7755–7760.

Kuipers, R. K., H. J. Joosten, W. J. van Berkel, N. G. Leferink, E. Rooijen, E. Ittmann, F. van Zimmeren et al. (2010). 3DM: Systematic analysis of heterogeneous superfamily data to discover protein functionalities. *Proteins* **78**(9): 2101–2113.

Kumar, L., G. Awasthi and B. Singh (2011a). Extremophiles: A novel source of industrially important enzymes. *Biotechnology* **10**(2): 121–135.

Kumar, P., D. M. Barrett, M. J. Delwiche and P. Stroeve (2011b). Pulsed electric field pretreatment of switchgrass and wood chip species for biofuel production. *Ind Eng Chem Res* **50**(19): 10996–11001.

Kurakake, M., N. Ide and T. Komaki (2007). Biological pretreatment with two bacterial strains for enzymatic hydrolysis of office paper. *Curr Microbiol* **54**(6): 424–428.

Kuwahara, M., J. K. Glenn, M. A. Morgan and M. H. Gold (1984). Separation and characterization of two extracelluar H_2O_2-dependent oxidases from ligninolytic cultures of *Phanerochaete chrysosporium. FEBS Lett* **169**(2): 247–250.

Ladisch, M. R., C. M. Ladisch and G. T. Tsao (1978). Cellulose to sugars: New path gives quantitative yield. *Science* **201**(4357): 743–745.

Lam, T. B. T., K. Iiyama and B. A. Stone (1992a). Cinnamic acid bridges between cell wall polymers in wheat and phalaris internodes. *Phytochemistry* **31**(4): 1179–1183.

Lam, T. B. T., K. Iiyama and B. A. Stone (1992b). Changes in phenolic acids from internode walls of wheat and phalaris during maturation. *Phytochemistry* **31**(8): 2655–2658.

Lam, T. B. T., K. Iiyama and B. A. Stone (1994). An approach to the estimation of ferulic acid bridges in unfractionated cell walls of wheat internodes. *Phytochemistry* **37**(2): 327–333.

Lam, T.-T., K. Iiyama and B. A. Stone (1990). Primary and secondary walls of grasses and other forage plants: Taxonomic and structural consideration. In *Microbial and Plant Opportunities to Improve Lignocellulose Uti- lization by Ruminants*, D. E. Akin, M. G. Ljundahl, R. J. Wilson and P. J. Hams (Eds.), pp. 43–69. New York: Elsevier.

Lamed, R., E. Setter and E. A. Bayer (1983a). Characterization of a cellulose-binding, cellulase-containing complex in *Clostridium thermocellum. J Bacteriol* **156**(2): 828–836.

Lamed, R., E. Setter, R. Kenig and E. A. Bayer (1983b). The cellulosome: A discrete cell surface organelle of *Clostridium thermocellum* which exhibits separate antigenic, cellulose-binding and various cellulolytic activities. *Biotechnol Bioeng Symp* **13**: 163–181.

Lan, W., C.-F. Liu and R.-C. Sun (2011). Fractionation of bagasse into cellulose, hemicelluloses, and lignin with ionic liquid treatment followed by alkaline extraction. *J Agric Food Chem* **59**(16): 8691–8701.

Landis, D. A., M. M. Gardiner, W. van der Werf and S. M. Swinton (2008). Increasing corn for biofuel production reduces biocontrol services in agricultural landscapes. *Proc Natl Acad Sci USA* **105**(51): 20552–20557.

Langston, J., T. Shaghasi, E. Abbate, F. Xu, E. Vlasenko and M. Sweeney (2011). Oxidoreductive cellulose depolymerization by the enzymes cellobiose dehydrogenase and glycoside hydrolase 61. *Appl Environ Microbiol* **77**: 7007–7015.

Lara-Marquez, A., M. G. Zavala-Paramo, E. Lopez-Romero and H. C. Camacho (2001). Biotechnological potential of pectinolytic complexes of fungi. *Biotechnol. Lett.* **33**: 859–868.

Larsbrink, J., A. Izumi, F. M. Ibatullin, A. Nakhai, H. J. Gilbert, G. J. Davies and H. Brumer (2011). Structural and enzymatic characterization of a glycoside hydrolase family 31 α-xylosidase from *Cellvibrio japonicus* involved in xyloglucan saccharification. *Biochem J* **436**(3): 567–580.

Laureano-Perez, L., F. Teymouri, H. Alizadeh and B. E. Dale (2005). Understanding factors that limit enzymatic hydrolysis of biomass: Characterization of pretreated corn stover. *Appl Biochem Biotechnol* **121– 124**: 1081–1099.

Lebbink, J. H. G., T. Kaper, P. Bron, J. van der Oost and W. M. de Vos (2000). Improving low-temperature catalysis in the hyperthermostable *Pyrococcus furiosus* β-glucosidase CelB by directed evolution. *Biochemistry* **39**(13): 3656–3665.

Lee, C. C., R. E. Kibblewhite, K. Wagschal, R. Li, G. H. Robertson and W. J. Orts (2012a). Isolation and characterization of a novel GH67 α-glucuronidase from a mixed culture. *J Ind Microbiol Biotechnol* **39**(8): 1245–1251.

Lee, H. L., C. K. Chang, W. Y. Jeng, A. H. Wang and P. H. Liang (2012b). Mutations in the substrate entrance region of β-glucosidase from *Trichoderma reesei* improve enzyme activity and thermostability. *Protein Eng Des Sel* **25**: 733–740.

Lee, J. (1997). Biological conversion of lignocellulosic biomass to ethanol. *J Biotechnol* **56**(1): 1–24.

Lee, R. C., R. A. Burton, M. Hrmova and G. B. Fincher (2001). Barley arabinoxylan arabinofuranohydrolases: Purification, characterization and determination of primary structures from cDNA clones. *Biochem J* **356**(Pt 1): 181–189.

Lee, R. C., M. Hrmova, R. A. Burton, J. Lahnstein and G. B. Fincher (2003). Bifunctional family 3 glycoside hydrolases from barley with alpha-L-arabinofuranosidase and beta-D-xylosidase activity. Characterization, primary structures, and COOH-terminal processing. *J Biol Chem* **278**(7): 5377–5387.

Lee, S. H., T. V. Doherty, R. J. Linhardt and J. S. Dordick (2009). Ionic liquid-mediated selective extraction of lignin from wood leading to enhanced enzymatic cellulose hydrolysis. *Biotechnol Bioeng* **102**(5): 1368–1376.

Lehmann, C., M. Bocola, W. R. Streit, R. Martinez and U. Schwaneberg (2014). Ionic liquid and deep eutectic solvent-activated CelA2 variants generated by directed evolution. *Appl Microbiol Biotechnol* **98**(12): 5775–5785.

Leibovitz, E. and P. Béguin (1996). A new type of cohesin domain that specifically binds the dockerin domain of the *Clostridium thermocellum* cellulosome-integrating protein CipA. *J Bacteriol* **178**(11): 3077–3084.

Leibovitz, E., H. Ohayon, P. Gounon and P. Béguin (1997). Characterization and subcellular localization of the *Clostridium thermocellum* scaffoldin dockerin binding protein SdbA. *J Bacteriology* **179**(8): 2519–2523.

Leisola, M. S. A., B. Schmidt, U. Thanei-Wyss and A. Fiechter (1985). Aromatic ring cleavage of veratryl alcohol by *Phanerochaete chrysosporium*. *FEBS Lett* **189**(2): 267–270.

Lemaire, M., H. Ohayon, P. Gounon, T. Fujino and P. Béguin (1995). OlpB, a new outer layer protein of *Clostridium thermocellum*, and binding of its S-layer-like domains to components of the cell envelope. *J Bacteriol* **177**(9): 2451–2459.

Léonard, R., M. Pabst, J. S. Bondili, G. Chambat, C. Veit, R. Strasser and F. Altmann (2008). Identification of an *Arabidopsis* gene encoding a GH95 alpha1,2-fucosidase active on xyloglucan oligo- and polysaccharides. *Phytochemistry* **69**(10): 1983–1988.

Leonowicz, A., K. Grzywnowicz and M. Malinowska (1979). Oxidative and demethylating activity of multiple forms of laccase from Pholiota mutabilis. *Acta Biochim Pol* **26**(4): 431–434.

Leung, D. W., E. Chen and D. V. Goeddel (1989). A method for random mutagenesis of a defined DNA segment using a modified polymerase chain reaction. *Technique* **1**: 11–15.

Levigne, S. V., M. C. Ralet, B. C. Quemener, B. N. Pollet, C. Lapierre and J. F. Thibault (2004). Isolation from sugar beet cell walls of arabinan oligosaccharides esterified by two ferulic acid monomers. *Plant Physiol* **134**(3): 1173–1180.

Levy, S., G. Maclachlan and L. A. Staehelin (1997). Xyloglucan sidechains modulate binding to cellulose during in vitro binding assays as predicted by conformational dynamics simulations. *Plant J* **11**(3): 373–386.

Levy, S., W. S. York, R. Stuike-Prill, B. Meyer and L. A. Staehelin (1991). Simulations of the static and dynamic molecular conformations of xyloglucan. The role of the fucosylated sidechain in surface-specific sidechain folding. *Plant J* **1**(2): 195–215.

Lewandowski, I., J. C. Clifton-Brown, J. M. O. Scurlock and W. Huisman (2000). Miscanthus: European experience with a novel energy crop. *Biomass Bioenergy* **19**(4): 209–227.

Li, A. and M. Khraisheh (2008). Municipal solid waste used as bioethanol sources and its related environmental impacts. *Int J Soil Sediment Water* **1**(1): 5–10.

Li, C., B. Knierim, C. Manisseri, R. Arora, H. V. Scheller, M. Auer, K. P. Vogel, B. A. Simmons and S. Singh (2010). Comparison of dilute acid and ionic liquid pretreatment of switchgrass: Biomass recalcitrance, delignification and enzymatic saccharification. *Bioresour Technol* **101**(13): 4900–4906.

Li, C., L. Wang, Z. Chen, Y. Li, R. Wang, X. luo, G. Cai, Y. Li, Q. Yu and J. Lu (2015). Ozonolysis pretreatment of maize stover: The interactive effect of sample particle size and moisture on ozonolysis process. *Bioresour Technol* **183**: 240–247.

Li, H., N.-J. Kim, M. Jiang, J. W. Kang and H. N. Chang (2009). Simultaneous saccharification and fermentation of lignocellulosic residues pretreated with phosphoric acid–acetone for bioethanol production. *Bioresour Technol* **100**(13): 3245–3251.

Li, H.-Q., W. Jiang, J.-X. Jia and J. Xu (2014). pH pre-corrected liquid hot water pretreatment on corn stover with high hemicellulose recovery and low inhibitors formation. *Bioresour Technol* **153**: 292–299.

Li, K., P. Azadi, R. Collins, J. Tolan, J. S. Kim and K. E. L. Eriksson (2000). Relationships between activities of xylanases and xylan structures. *Enzyme Microb Technol* **27**(1–2): 89–94.

Li, S., X. Zhang and J. M. Andresen (2012). Production of fermentable sugars from enzymatic hydrolysis of pretreated municipal solid waste after autoclave process. *Fuel* **92**: 84–88.

Li, W., N. Sun, B. Stoner, X. Jiang, X. Lu and R. D. Rogers (2011). Rapid dissolution of lignocellulosic biomass in ionic liquids using temperatures above the glass transition of lignin. *Green Chem* **13**: 2038–2047.

Li, X., W. T. Beeson, C. M. Phillips, M. A. Marletta and J. H. D. Cate (2012). Structural basis for substrate targeting and catalysis by fungal polysaccharide monooxygenases. *Structure* **20**(6): 1051–1061.

Li, X., J.-K. Weng and C. Chapple (2008a). Improvement of biomass through lignin modification. *Plant J* **54**(4): 569–581.

Li, X. L., H. Chen and L. G. Ljungdahl (1997). Two cellulases, CelA and CelC, from the polycentric anaerobic fungus *Orpinomyces* strain PC-2 contain N-terminal docking domains for a cellulase-hemicellulase complex. *Appl Environ Microbiol* **63**(12): 4721–4728.

Li, X.-L., C. D. Skory, M. A. Cotta, V. Puchart and P. Biely (2008b). Novel family of carbohydrate esterases, based on identification of the *Hypocrea jecorina* acetyl esterase gene. *Appl Environ Microbiol* **74**(24): 7482–7489.

Li, X.-L., S. Špániková, R. P. de Vries and P. Biely (2007a). Identification of genes encoding microbial glucuronoyl esterases. *FEBS Lett* **581**(21): 4029–4035.

Li, Y., D. C. Irwin and D. B. Wilson (2007b). Processivity, substrate binding, and mechanism of cellulose hydrolysis by *Thermobifida fusca* Cel9A. *Appl Environ Microbiol* **73**(10): 3165–3172.

Liang, C., M. Fioroni, F. Rodriguez-Ropero, Y. Xue, U. Schwaneberg and Y. Ma (2011). Directed evolution of a thermophilic endoglucanase (Cel5A) into highly active Cel5A variants with an expanded temperature profile. *J Biotechnol* **154**(1): 46–53.

Liebgott, P. P., A. L. de Lacey, B. Burlat, L. Cournac, P. Richaud, M. Brugna, V. M. Fernandez et al. (2010). Original design of an oxygen-tolerant [NiFe] hydrogenase: Major effect of a valine-to-cysteine mutation near the active site. *J Am Chem Soc* **133**(4): 986–997.

Liers, C., C. Bobeth, M. Pecyna, R. Ullrich and M. Hofrichter (2010). DyP-like peroxidases of the jelly fungus *Auricularia auricula-judae* oxidize nonphenolic lignin model compounds and high-redox potential dyes. *Appl Microbiol Biotechnol* **85**(6): 1869–1879.

Lin, L., X. Meng, P. Liu, Y. Hong, G. Wu, X. Huang, C. Li, J. Dong, L. Xiao and Z. Liu (2009). Improved catalytic efficiency of endo-beta-1,4-glucanase from *Bacillus subtilis* BME-15 by directed evolution. *Appl Microbiol Biotechnol* **82**(4): 671–679.

Linder, Å., J. P. Roubroeks and P. Gatenholm (2005). Effect of ozonation on assembly of xylans. *Holzforschung* **57**(5): 496–502.

Lisk, D. J. (1988). Environmental implications of incineration of municipal solid waste and ash disposal. *Sci Total Environ* **74**: 39–66.

Liu, H., L. Zhu, M. Bocola, N. Chen, A. C. Spiess and U. Schwaneberg (2013a). Directed laccase evolution for improved ionic liquid resistance. *Green Chem* **15**(5): 1348–1355.

Liu, M., J. Gu, W. Xie and H. Yu (2013b). Directed co-evolution of an endoglucanase and a β-glucosidase in *Escherichia coli* by a novel high-throughput screening method. *Chem Commun* **49**: 7219–7221.

Liu, W., X. Z. Zhang, Z. Zhang and Y. H. Zhang (2010). Engineering of Clostridium phytofermentans endoglucanase Cel5A for improved thermostability. *Appl Environ Microbiol* **76**(14): 4914–4917.

Liu, Y.-S., J. Baker, Y. Zeng, M. Himmel, T. Haas and S.-Y. Ding (2011). Cellobiohydrolase hydrolyzes crystalline cellulose on hydrophobic faces. *J Biol Chem* **286**(13): 11195–11201.

Lombard, V., H. Golaconda Ramulu, E. Drula, P. M. Couthino and B. Henrissat (2014). The carbohydrate-active enzymes database (CAZy) in 2013. *Nucleic Acids Res* **42**(D1): D490–D495.

Lora, J. H. and M. Wayman (1978). Delignification of hardwoods by auto-hydrolysis and extraction. *Tappi J* **61**(6): 47–50.

Lou, X. F. and J. Nair (2009). The impact of landfilling and composting on greenhouse gas emissions—a review. *Bioresour Technol* **100**(16): 3792–3798.

Lundqvist, J., A. Jacobs, M. Palm, G. Zacchi, O. Dahlman and H. Stålbrand (2003). Characterization of galactoglucomannan extracted from spruce (*Picea abies*) by heat-fractionation at different conditions. *Carbohydr Polym* **51**(2): 203–211.

Lunetta, J. M. and D. Pappagianis (2014). Identification, molecular characterization, and expression analysis of a DOMON-like type 9 carbohydrate-binding module domain-containing protein of *Coccidioides posadasii*. *Med Mycol* **52**(6): 591–609.

Luterbacher, J. S., J. Y. Parlange and L. P. Walker (2013). A pore-hindered diffusion and reaction model can help explain the importance of pore size distribution in enzymatic hydrolysis of biomass. *Biotechnol Bioeng* **110**(1): 127–136.

Lutzen, N. W., M. H. Nielsen, K. M. Oxiboell, M. Schiilein and B. S. Olessen (1983). Cellulase and their application in the conversion of lignocellulose to fermentable sugar. *Philos Trans R Soc London* **300**: 283.

Lynd, L. R., J. H. Cushman, R. J. Nichols and C. E. Wyman (1991). Fuel ethanol from cellulosic biomass. *Science* **251**(4999): 1318–1323.

Lynd, L. R., R. T. Elamder and C. E. Wyman (1996). Likely features and costs of mature biomass ethanol technology. *Appl Biochem Biotechnol* **57– 58**(1): 741–761.

Lynd, L. R., P. J. Weimer, W. H. van Zyl and I. S. Pretorius (2002). Microbial cellulose utilization: Fundamentals and biotechnology. *Microbiol Mol Biol Rev* **66**(3): 506–577.

Lytle, B. L., B. F. Volkman, W. M. Westler, M. P. Heckman and J. H. Wu (2001). Solution structure of a type I dockerin domain, a novel prokaryotic, extracellular calcium-binding domain. *J Mol Biol* **307**(3): 745–753.

Lytle, B. L., B. F. Volkman, W. M. Westler and J. H. D. Wu (2000). Secondary structure and calcium-induced folding of the *Clostridium thermocellum* dockerin domain determined by NMR spectroscopy. *Arch Biochem Biophys* **379**(2): 237–244.

Lytle, B. L., W. M. Westler and J. H. D. Wu (1999). Molecular assembly of the *Clostridium thermocellum* cellulosome. In *Genetics, Biochemistry, and Ecology of Cellulose Degradation*, K. Ohmiya, K. Hayashi, K. Sakka et al. (Eds.), pp. 444–449. Tokyo, Japan: Uni Publishers.

Ma, F., N. Yang, C. Xu, H. Yu, J. Wu and X. Zhang (2010). Combination of biological pretreatment with mild acid pretreatment for enzymatic hydrolysis and ethanol production from water hyacinth. *Bioresour Technol* **101**: 9600–9604.

Mackie, K. L., H. H. Brownell, K. L. West and J. N. Saddler (1985). Effect of sulphur dioxide and sulphuric acid on steam explosion of aspenwood. *J Wood Chem Technol* **5**(3): 405–425.

Madigan, M. T., J. M. Martinko and J. Parker (2000). Nutrition and metabolism. In *Brock Biology of Microbiology*, M. T. Madigan, J. M. Martinko and J. Parker (Eds.). Upper Saddle River, NJ: Prentice-Hall.

Mai, V., J. Wiegel and W. W. Lorenz (2000). Cloning, sequencing, and characterization of the bifunctional xylosidase-arabinosidase from the anaerobic thermophile thermoanaerobacter ethanolicus. *Gene* **247**(1–2): 137–143.

Maiorella, B. I. (1983). Ethanol industrial chemicals. In *Biochem Fuels*, pp. 861–914. Oxford: Pergamon Press.

Maiorella, B. L. (1985). Ethanol. In *Comprehensive Biotechnology*, M. Moo-Young (Ed.), pp. 861–914. Oxford: Pergamon Press.

Malarczyk, E., J. Kochmanska-Rdest and A. Jarosz-Wilkolazka (2009). Influence of very low doses of mediators on fungal laccase activity—Nonlinearity beyond imagination. *Nonlinear Biomed Phys* **3**: 10.

Mandels, M. and E. T. Reese (1965). Inhibition of cellulase. *Ann Rev Phytopathol* **3**: 85–102.

Mandels, M. and E. T. Reese (1999). Fungal cellulases and the microbial decomposition of cellulosic fabric. *J Ind Microbiol Biotechnol* **5**(4–5): 225–240.

Marchessault, R. H., H. Holava and T. E. Timell (1963). Correlation of chemical composition and optical rotation of native xylans. *Can J Chem* **41**(6): 1612–1618.

Marchessault, R. H. and A. Sarko (1967). X-ray structure opolysaccharides. *Adv Carbohydr Chem Biochem* **22**: 421–482.

Marcus, S. E., Y. Verhertbruggen, C. Herve, J. J. Ordaz-Ortiz, V. Farkas, H. L. Pedersen, W. G. Willats and J. P. Knox (2008). Pectic homogalacturonan masks abundant sets of xyloglucan epitopes in plant cell walls. *BMC Plant Biol* **8**: 60.

Margolles-Clark, E., M. Saloheimo, M. Siika-aho and M. Penttilä (1996a). The α-glucuronidase-encoding gene of *Trichoderma reesei*. *Gene* **172**(1): 171–172.

Margolles-Clark, E., M. Tenkanen, E. Luonteri and M. Penttilä (1996b). Three α-galactosidase genes of *Trichoderma reesei* cloned by expression in yeast. *Eur J Biochem* **240**(1): 104–111.

Margolles-Clark, E., M. Tenkanen, T. Nakari-Setälä and M. Penttilä (1996c). Cloning of genes encoding alpha-L-arabinofuranosidase and beta-xylosidase from *Trichoderma reesei* by expression in *Saccharomyces cerevisiae*. *Appl Environ Microbiol* **62**(10): 3840–3846.

Margolles-Clark, E., M. Tenkanen, H. Söderlund and M. Penttilä (1996d). Acetyl xylan esterase from *Trichoderma reesei* contains an active-site serine residue and a cellulose-binding domain. *Eur J Biochem* **237**(3): 553–560.

Mark, H. and A. V. Tobolsky (1950). *Physical Chemistry of High Polymeric Systems*. New York: Interscience.

Markovic, O. and S. Janecek (2001). Pectin degrading glycoside hydrolases of family 28: Sequence-structural features, specificities and evolution. *Protein Eng* **14**(9): 615–631.

Martín, C., H. B. Klinke and A. B. Thomsen (2007). Wet oxidation as a pretreatment method for enhancing the enzymatic convertibility of sugarcane bagasse. *Enzyme Microb Technol* **40**(3): 426–432.

Martin, O. (2009). *Dilute sulfuric acid pretreatment of Switchgrass in microwave reactor for biofuel conversion: An investigation of yields, kinetics*. Doctor of Philosophy, Virginia Commonwealth University.

Martinez, D., R. Berka, B. Henrissat, M. Saloheimo, M. Arvas, S. Baker, J. Chapman, O. Chertkov, P. Coutinho and D. Cullen (2008). Genome sequencing and analysis of the biomass-degrading fungus *Trichoderma reesei* (syn. *Hypocrea jecorina*). *Nat Biotechnol* **26**(5): 553–560.

Martinez, M. J., B. Böckle, S. Camarero, F. Guillén and A. T. Martinez (1996a). MnP isoenzymes produced by two Pleurotus species in liquid culture and during wheat-straw solid-state fermentation. In *Enzymes for Pulp and Paper Processing*, Vol. 655, pp. 183–196. Washington, DC: American Chemical Society.

Martinez, M. J., F. J. Ruiz-Duenas, F. Guillen and A. T. Martinez (1996b). Purification and catalytic properties of two manganese peroxidase isoenzymes from *Pleurotus eryngii*. *Eur J Biochem* **237**(2): 424–432.

Mateo, C., J. M. Palomo, G. Fernandez-Lorente, J. M. Guisan and R. Fernandez-Lafuente (2007). Improvement of enzyme activity, stability and selectivity via immobilization techniques. *Enzyme Microb Technol* **40**(6): 1451–1463.

Matheson, N. K. (1990). 11-Mannose-based Polysaccharides. In *Methods in Plant Biochemistry*, P. M. Dey (Ed.), Vol. 2, pp. 371–413. London: Academic Press.

Matsumura, I. (2000). Accelerating the discovery, modification and commercialization of enzyme applications. *IBC's Fifth Annual World Congress on Enzyme Technologies*, Las Vegas, CA.

Matsuo, M. and T. Yasui (1984). Purification and some properties of beta-xylosidase from *Trichoderma viride*. *Agric Biol Chem* **48**(7): 1845–1852.

Matte, A. and C. W. Forsberg (1992). Purification, characterization, and mode of action of endoxylanases 1 and 2 from *Fibrobacter succinogenes* S85. *Appl Environ Microbiol* **58**(1): 157–168.

Maurya, D. P., A. Singla and S. Negi (2015). An overview of key pretreatment processes for biological conversion of lignocellulosic biomass to bioethanol. *Biotechnology* **3**: 1–13.

Maurya, D. P., D. Singh, D. Pratap and J. P. Maurya (2012). Optimization of solid state fermentation conditions for the production of cellulase by *Trichoderma reesei. J Environ Biol* **33**(1): 5–8.

May, O., P. T. Nguyen and F. H. Arnold (2000). Inverting enantioselectivity by directed evolution of hydantoinase for improved production of L-methionine. *Nat Biotechnol* **18**(3): 317–320.

Mayans, O., M. Scott, I. Connerton, T. Gravesen, J. Benen, J. Visser, R. Pickersgill and J. Jenkins (1997). Two crystal structures of pectin lyase A from *Aspergillus* reveal a pH driven conformational change and striking divergence in the substrate-binding clefts of pectin and pectate lyases. *Structure* **5**(5): 677–689.

Mayer, F., M. P. Coughlan, Y. Mori and L. G. Ljungdahl (1987). Macromolecular organization of the cellulolytic enzyme complex of *Clostridium thermocellum* as revealed by electron microscopy. *Appl Environ Microbiol* **53**(12): 2785–2792.

McCarthy, J. K., A. Uzelac, D. F. Davis and D. E. Eveleigh (2004). Improved catalytic efficiency and active Sste modification of 1,4-β-D-glucan glucohydrolase a from *Thermotoga neapolitana* by directed evolution. *J Biol Chem* **279**(12): 11495–11502.

McDonald, R. I., J. Fargione, J. Kiesecker, W. M. Miller and J. Powell (2009). Energy sprawl or energy efficiency: Climate policy impacts on natural habitat for the United States of America. *PLoS One* **4**(8): e6802.

McGhee, J. E., M. E. Carr and G. S. Julian (1984). Continuous bioconversion of starch to ethanol by calcium-alginate immobilized enzymes and yeasts. *Cereal Chem* **61**(5): 446–449.

McGinnis, G. D., W. W. Wilson and C. E. Mullen (1983). Biomass pretreatment with water and high-pressure oxygen. The wet-oxidation process. *Ind Eng Chem Prod Res Dev* **22**(2): 352–357.

McLaughlin, S. B., J. Bouton, D. Bransby, B. Conger, W. Ocumpaugh, D. Parrish, C. Taliaferro, K. P. Vogel and S. Wullschleger (1999). Developing switchgrass as a bioenergy feedstock. In *Perspectives on New Crops and New Uses*, J. J. Alexandria (Ed.), pp. 282–299. Alexandria, VA: ASHS Press.

McMillan, J. (1994). Pretreatment of lignocellulosic biomass. In *Enzymatic Conversion of Biomass for Fuels Production*, M. Himmel, J. Baker and R. Overend (Ed.), Vol. 566, pp. 292–324. Washington, DC: American Chemical Society.

McNeil, M., A. G. Darvill and P. Albersheim (1980). Structure of plant cell walls: X. Rhamnogalacturonan I, a structurally complex pectic polysaccharide in the walls of suspension-cultured sycamore cells. *Plant Physiol* **66**(6): 1128–1134.

Meier, H. and J. S. G. Reid (1982). Reserve polysaccharides other than starch in higher plants. In *Encyclopedia of Plant*, F. A. Loewus and W. Tanner (Eds.), Vol. 13A, pp. 418–471. Berlin, Germany: Springer.

Mel'nik, M. S., D. V. Kapkov, M. A. Mogutov, M. L. Rabinovich and A. A. Klesov (1989). A new type of *Clostridium thermocellum* endoglucanase produced by the recombinant strain of *E. coli*. Some properties and identification in donor cells. *Biokhimiia* **54**: 387–395.

Mel'nik, M. S., M. L. Rabinovich and Y. V. Voznyi (1991). Cellobiohydrolase of *Clostridium thermocellum* produced by a recombinant *E. coli* strain. *Biokhimiya* **56**: 1787–1797.

Mello, B. L. and I. Polikarpov (2014). Family 1 carbohydrate binding-modules enhance saccharification rates. *AMB Express* **4**: 36.

Melzer, S., C. Sonnendecker, C. Föllner and W. Zimmermann (2015). Stepwise error-prone PCR and DNA shuffling changed the pH activity range and product specificity of the cyclodextrin glucanotransferase from an alkaliphilic *Bacillus* sp. *FEBS Open Bio* **5**: 528–534.

Menon, V. and M. Rao (2012). Trends in bioconversion of lignocellulose: Biofuels, platform chemicals & biorefinery concept. *Prog Energy Combust Sci* **38**(4): 522–550.

Mercer, D. K., M. Iqbal, P. Miller and A. J. McCarthy (1996). Screening actinomycetes for extracellular peroxidase activity. *Appl Environ Microbiol* **62**(6): 2186–2190.

Merino, S. and J. Cherry (2007a). Progress and challenges in enzyme development for biomass utilization. In *Biofuels*, L. Olsson (Ed.), Vol. 108, pp. 95–120. Berlin, Germany: Springer.

Merino, S. T. and J. Cherry (2007b). Progress and challenges in enzyme development for biomass utilization. In *Advances in Biochemical Engineering/ Biotechnology*, T. Scheper, S. Belkin, P. M. Doran et al. (Eds.), Vol. 108, pp. 95–120. Berlin, Germany: Springer.

Messerschmidt, A. (1997). *Multi-copper Oxidases*. Singapore: World Scientific Pub Co.

Mester, T., K. Ambert-Balay, S. Ciofi-Baffoni, L. Banci, A. D. Jones and M. Tien (2001). Oxidation of a tetrameric nonphenolic lignin model compound by lignin peroxidase. *J Biol Chem* **276**(25): 22985–22990.

Metz, B., V. Seidl-Seiboth, T. Haarmann, A. Kopchinskiy, P. Lorenz, B. Seiboth and C. P. Kubicek (2011). Expression of biomass-degrading enzymes is a major event during conidium development in *Trichoderma reesei*. *Eukaryotic Cell* **10**(11): 1527–1535.

Meyssami, B., M. O. Balaban and A. A. Teixeira (1992). Prediction of pH in model systems pressurized with carbon-dioxide. *Biotechnol Prog* **8**: 149–154.

Michael, W., H. Jeongwoo, B. D. Jennifer, C. Hao and E. Amgad (2012). Well-to-wheels energy use and greenhouse gas emissions of ethanol from corn, sugarcane and cellulosic biomass for US use. *Environ Res Lett* **7**(4): 045905.

Mielenz, J. R. (2001). Ethanol production from biomass: Technology and commercialization status. *Curr Opin Microbiol* **4**: 324–329.

Mihoc, A. and D. Kluepfel (1990). Purification and characterization of a β-glucosidase from Streptomyces lividans 66. *Can J Microbiol* **36**(1): 53–56.

Miller, R. B. (1999). Structure of wood. In *Wood Handbook: Wood as an Engineering Material*, Vol. 113. Madison, WI: USDA Forest Service, Forest Products Laboratory.

Millett, M. A., A. J. Baker and L. D. Satter (1976). Physical and chemical pretreatments for enhancing cellulose saccharification. *Biotechnol Bioeng Symp* **6**: 125–153.

Millett, M. A., M. J. Effland and D. F. Caulfield (1979). Influence of fine grinding on the hydrolysis of cellulosic materials—Acid vs. enzymatic. In *Hydrolysis of Cellulose: Mechanisms of Enzymatic and Acid Catalysis*, Vol. 181, pp. 71–89. Washington, DC: American Chemical Society.

Millis, C. D., D. Cai, M. T. Stankovich and M. Tien (1989). Oxidation-reduction potentials and ionization states of extracellular peroxidases from the lignin-degrading fungus *Phanerochaete chrysosporium*. *Biochemistry* **28**(21): 8484–8489.

Mills, D. R., R. L. Peterson and S. Spiegelman (1967). An extracellular Darwinian experiment with a self-duplicating nucleic acid molecule. *Proc Natl Acad Sci USA* **58**(1): 217–224.

Mingos, D. M. P. and D. R. Baghurst (1991). Tilden Lecture. Applications of microwave dielectric heating effects to synthetic problems in chemistry. *Chem Soc Rev* **20**: 1–47.

Mirahmadi, K., M. M. Kabir, A. Jeihanipour, K. Karimi and M. J. Taherzadeh (2010). Alkaline pretreatment of spruce and birch to improve bioethanol and biogas production. *Bioresources* **5**(2): 928–938.

Miras, I., F. Schaeffer, P. Béguin and P. M. Alzari (2002). Mapping by site-directed mutagenesis of the region responsible for cohesin–dockerin interaction on the surface of the seventh cohesin domain of *Clostridium thermocellum* CipA. *Biochemistry* **41**(7): 2115–2119.

Mitsuzawa, S., H. Kagawa, Y. Li, S. L. Chan, C. D. Paavola and J. D. Trent (2009). The rosettazyme: A synthetic cellulosome. *J Biotechnol* **143**(2): 139–144.

Mittal, A., R. Katahira, M. E. Himmel and D. K. Johnson (2011). Effects of alkaline or liquid-ammonia treatment on crystalline cellulose: Changes in crystalline structure and effects on enzymatic digestibility. *Biotechnol Biofuels* **4**: 41.

Mohamed, S. A., N. M. Farid, E. N. Hossiny and R. I. Bassuiny (2006). Biochemical characterization of an extracellular polygalacturonase from *Trichoderma harzianum*. *J Biotechnol* **127**(1): 54–64.

Mohanram, S., D. Amat, J. Choudhary, A. Arora and L. Nain (2013). Novel perspectives for evolving enzyme cocktails for lignocellulose hydrolysis in biorefineries. *Sustain Chem Process* **1**: 15.

Mohnen, D. (2008). Pectin structure and biosynthesis. *Curr Opin Plant Biol* **11**(3): 266–277.

Molinier, A.-L., M. Nouailler, O. Valette, C. Tardif, V. Receveur-Brechot and H.-P. Fierobe (2011). Synergy, structure, and conformational flexibility of hybrid cellulosomes displaying various inter-cohesin linkers. *J Mol Biol* **405**: 143–157.

Monlau, F., Q. Aemig, A. Barakat, J. P. Steyer and H. Carrère (2013). Application of optimized alkaline pretreatment for enhancing the anaerobic digestion of different sunflower stalks varieties. *Environ Technol* **34**(13–14): 2155–2162.

Mood, S. H., A. H. Golfeshan, M. Tabatabaei, G. S. Jouzani, G. H. Najafi, M. Gholami and M. Ardjmand (2013). Lignocellulosic biomass to bioethanol, a comprehensive review with a focus on pretreatment. *Renew Sustainable Energy Rev* **27**: 77–93.

Moore, J. C. and F. H. Arnold (1996). Directed evolution of a para-nitrobenzyl esterase for aqueous-organic solvents. *Nat Biotechnol* **14**(4): 458–467.

Morag, E., E. A. Bayer and R. Lamed (1990). Relationship of cellulosomal and noncellulosomal xylanases of *Clostridium thermocellum* to cellulose-degrading enzymes. *J Bacteriol* **172**(10): 6098–6105.

Morag, E., A. Lapidot, D. Govorko, R. Lamed, M. Wilchek, E. A. Bayer and Y. Shoham (1995). Expression, purification, and characterization of the cellulose-binding domain of the scaffoldin subunit from the cellulosome of *Clostridium thermocellum*. *Appl Environ Microbiol* **61**(5): 1980–1986.

Moran, F., S. Nasuno and M. P. Starr (1968). Extracellular and intracellular polyglacturonic acid trans-eliminases of *Erwinia carotovora*. *Arch Biochem Biophys* **123**(2): 298–306.

Moreira, L. R. S. and E. X. F. Filho (2008). An overview of mannan structure and mannan-degrading enzyme systems. *Appl Microbiol Biotechnol* **79**(2): 165–178.

Mori, Y., S. Ozasa, M. Kitaoka, S. Noda, T. Tanaka, H. Ichinose and N. Kamiya (2013). Aligning an endoglucanase Cel5A from *Thermobifida fusca* on a DNA scaffold: Potent design of an artificial cellulosome. *Chem Commun R Soc Chem* **49**: 6971–6973.

Morias, S., E. Morag, Y. Barak, D. Goldman, Y. Hadar, R. Lamed, Y. Shoham, D. B. Wilson and E. A. Bayer (2012). Deconstruction of lignocellulose into soluble sugars by native and designer cellulosomes. *mBio* **3**(5): 1–11.

Morley, K. L. and R. J. Kazlauskas (2005). Improving enzyme properties: When are closer mutations better? *Trends Biotechnol* **23**(5): 231–237.

Morpeth, F. F. (1985). Some properties of cellobiose oxidase from the white-rot fungus *Sporotrichum pulverulentum*. *Biochem J* **228**(3): 557–564.

Mort, A. J., F. Qiu and N. O. Maness (1993). Determination of the pattern of methyl esterification in pectin. Distribution of contiguous nonesterified residues. *Carbohydr Res* **247**: 21–35.

Moser, F., D. Irwin, S. Chen and D. Wilson (2008). Regulation and characterization of *Thermobifida fusca* carbohydrate-binding module proteins E7 and E8. *Biotechnol Bioeng* **100**: 1066–1077.

Mosier, N., R. Hendrickson, N. Ho, M. Sedlak and M. R. Ladisch (2005a). Optimization of pH controlled liquid hot water pretreatment of corn stover. *Bioresour Technol* **96**(8): 1986–1993.

Mosier, N., C. Wyman, B. Dale, R. Elander, Y. Y. Lee, M. Holtzapple and M. Ladisch (2005b). Features of promising technologies for pretreatment of lignocellulosic biomass. *Bioresour Technol* **96**(6): 673–686.

Mosier, N. S., R. Hendrickson, M. Brewer, N. Ho, M. Sedlak, R. Dreshel, G. Welch, B. S. Dien, A. Aden and M. R. Ladisch (2005c). Industrial scale-up of pH-controlled liquid hot water pretreatment of corn fiber for fuel ethanol production. *Appl Biochem Biotechnol* **125**(2): 77–97.

Mosier, N. S., R. Hendrickson and R. Dreschel (2003a). Principles and economics of pretreating cellulose in water for ethanol production. *BIOT Division, 225th American Chemical Society Meeting*, New Orleans, CA.

Mosier, N. S., R. Hendrickson and G. Welch (2003b). Corn fiber pretreatment scale-up and evaluation in an industrial corn to ethanol facility. *Proceedings of the 25th Symposium on Biotechnology for Fuels and Chemicals*, May 7, Breckenridge, CO.

Mosolova, T. P., S. V. Kalyuzhnyi, S. D. Varfolomeyev and G. A. Velikodvorskaya (1993). Purification and properties of *Clostridium thermocellum* endoglucanase 5 produced in Escherichia coli. *Appl Biochem Biotechnol* **42**(1): 9–18.

Moxley, G., Z. Zhu and Y. H. Zhang (2008). Efficient sugar release by the cellulose solvent-based lignocellulose fractionation technology and enzymatic cellulose hydrolysis. *J Agric Food Chem* **56**(17): 7885–7890.

Muheim, A., R. Waldner, D. Sanglard, J. Reiser, H. E. Schoemaker and M. S. Leisola (1991). Purification and properties of an aryl-alcohol dehydrogenase from the white-rot fungus *Phanerochaete chrysosporium*. *Eur J Biochem* **195**(2): 369–375.

Muller, D. (1928). Studies on the new enzyme glucose oxidase. I. *Biochem Z* **199**: 136–170.

Muller, D. (1936). Glucose oxidase. *Ergebn Enzymforsch* **5**: 259.

Murashima, K., A. Kosugi and R. H. Doi (2002). Thermostabilization of cellulosomal endoglucanase EngB from *Clostridium cellulovorans* by *in vitro* DNA recombination with non-cellulosomal endoglucanase EngD. *Mol Microbiol* **45**(3): 617–626.

Murugesan, K., M. Arulmani, I. H. Nam, Y. M. Kim, Y. S. Chang and P. T. Kalaichelvan (2006). Purification and characterization of laccase produced by a white rot fungus *Pleurotus sajor-caju* under submerged culture condition and its potential in decolorization of azo dyes. *Appl Microbiol Biotechnol* **72**(5): 939–946.

Mutter, M., G. Beldman, S. M. Pitson, H. A. Schols and A. G. Voragen (1998a). Rhamnogalacturonan α-D-galactopyranosyluronohydrolase. An enzyme that specifically removes the terminal nonreducing galacturonosyl residue in rhamnogalacturonan regions of pectin. *Plant Physiol* **117**: 153–163.

Mutter, M., G. Beldman, H. A. Schols and A. G. Voragen (1994). Rhamnogalacturonan alpha-L-rhamnopyranohydrolase. A novel enzyme specific for the terminal nonreducing rhamnosyl unit in rhamnogalacturonan regions of pectin. *Plant Physiol* **106**(1): 241–250.

Mutter, M., I. J. Colquhoun, G. Beldman, H. A. Schols, E. J. Bakx and A. G. Voragen (1998b). Characterization of recombinant rhamnogalacturonan alpha-L-rhamnopyranosyl-(1,4)-alpha-D-galactopyranosyluronide lyase from *Aspergillus* aculeatus. An enzyme that fragments rhamnogalacturonan I regions of pectin. *Plant Physiol* **117**(1): 141–152.

Mutter, M., I. J. Colquhoun, H. A. Schols, G. Beldman and A. G. Voragen (1996). Rhamnogalacturonase B from *Aspergillus aculeatus* is a rhamnogalacturonan alpha-L-rhamnopyranosyl-(1-->4)-alpha-D-galactopyranosyluronide lyase. *Plant Physiol* **110**(1): 73–77.

Nagy, T., R. B. Tunnicliffe, L. D. Higgins, C. Walters, H. J. Gilbert and M. Williamson (2007). Characterization of a double dockerin from the cellulosome of the anaerobic fungus *Piromyces* equi. *J Mol Biol* **373**(3): 612–622.

Naidu, S. L. and S. P. Long (2004). Potential mechanisms of low-temperature tolerance of C4 photosynthesis in *Miscanthus* × *giganteus*: An in vivo analysis. *Planta* **220**: 145–155.

Nakamura, A., H. Furuta, H. Maeda, T. Takao and Y. Nagamatsu (2002). Structural studies by stepwise enzymatic degradation of the main backbone of soybean soluble polysaccharides consisting of galacturonan and rhamnogalacturonan. *Biosci Biotechnol Biochem* **66**: 1301–1313.

Nakazawa, H., K. Okada, T. Onodera, W. Ogasawara, H. Okada and Y. Morikawa (2009). Directed evolution of endoglucanase III (Cel12A) from *Trichoderma reesei*. *Appl Microbiol Biotechnol* **83**(4): 649–657.

Narayanaswamy, N., A. Faik, D. J. Goetz and T. Gu (2011). Supercritical carbon dioxide pretreatment of corn stover and switchgrass for lignocellulosic ethanol production. *Bioresour Technol* **102**: 6995–7000.

Navarro, D., M. Couturier, G. G. da Silva, J. G. Berrin, X. Rouau, M. Asther and C. Bignon (2010). Automated assay for screening the enzymatic release of reducing sugars from micronized biomass. *Microb Cell Fact* **9**: 58.

Neas, E. D. and M. J. Collins (1988). *Introduction to Microwave Sample Preparation*. Washington, DC: ACS Professional Reference Book.

Ness, J. E., M. Welch, L. Giver, M. Bueno, J. R. Cherry, T. V. Borchert, W. P. Stemmer and J. Minshull (1999). DNA shuffling of subgenomic sequences of subtilisin. *Nat Biotechnol* **17**(9): 893–896.

Ng, T.-K., L. R. Gahan, G. Schenk and D. L. Ollis (2015). Altering the substrate specificity of methyl parathion hydrolase with directed evolution. *Arch Biochem Biophys.* **573**: 59–68.

Ng, T. K. and J. G. Zeikus (1981). Purification and characterization of an endoglucanase (1,4-beta-D-glucan glucanohydrolase) from *Clostridium thermocellum*.
Biochem J **199**(2): 341.

Nichols, N. N. and R. J. Bothast (2008). Production of ethanol from grain. In *Genetic
Improvement of Bioenergy Crops*, W. Vermerris (Ed.), pp. 75–88. New York:
Springer.

Nigam, J. N. (2001). Ethanol production from wheat straw hemicellulose hydrolysate by *Pichia stipitis. J Biotechnol* **87**(1): 17–27.

Nigam, P. S. and A. Singh (2011). Production of liquid biofuels from renewable
resources. *Prog Energy Combust Sci* **37**(1): 52–68.

Ninomiya, K., T. Yamauchi, C. Ogino, N. Shimizu and K. Takahashi (2014).
Microwave pretreatment of lignocellulosic material in cholinium ionic
liquid for efficient enzymatic saccharification. *Biochem Eng J* **90**: 90–95.

Nishiyam, Y., P. Langan, M. Wada and V. T. Forsyth (2010). Looking at hydrogen
bonds in cellulose. *Acta Crystallogr D Biol Crystallogr* **66**(11): 1172–1177.

NL Agency (2013). Bamboo: Analyzing the potential of bamboo feedstock for the
biobased economy. Utrecht, the Netherlands, p. 36.

Normark, M., S. Winestrand, T. A. Lestander and L. J. Jönsson (2014). Analysis,
pretreatment and enzymatic saccharification of different fractions of Scots
pine. *BMC Biotechnol* **14**: 20.

Northcote, D. H. (1972). Chemistry of the plant cell wall. *Annu Rev Plant Physiol*
23(1): 113–132.

Notenboom, V., A. B. Boraston, P. Chiu, A. C. Freelove, D. G. Kilburn and D. R.
Rose (2001). Recognition of cello-oligosaccharides by a family 17 carbohydrate-binding module: An X-ray crystallographic, thermodynamic and
mutagenic study. *J Mol Biol* **314**(4): 797–806.

Nousiainen, P., P. Maijala, A. Hatakka, A. T. Martínez and J. Sipilä (2009). Syringyl-
type simple plant phenolics as mediating oxidants in laccase catalyzed degradation of lignocellulosic materials: Model compound studies 10th EWLP,
Stockholm, Sweden, August 25–28, 2008. *Holzforschung* **63**(6): 699–704.

Nurizzo, D., T. Nagy, H. J. Gilbert and G. J. Davies (2002). The structural basis for
catalysis and specificity of the *Pseudomonas cellulosa* alpha-glucuronidase,
GlcA67A. *Structure* **10**(4): 547–556.

O'Neill, M. A., T. Ishii, P. Albersheim and A. G. Darvill (2004). Rhamnogalacturonan
II: Structure and function of a borate cross-linked cell wall pectic polysaccharide. *Annu Rev Plant Biol* **55**: 109–139.

O'Neill, M. A. and W. S. York (2003). The composition and structure of plant primary cell walls. In *The Plant Cell Wall*, J. K. C. Rose (Ed.), pp. 1–54. Ithaca:
Blackwell Publishing/CRC Press.

ØBro, J., J. Harholt, H. V. Scheller and C. Orfila (2004). Rhamnogalacturonan I
in *Solanum tuberosum* tubers contains complex arabinogalactan structures.
Phytochemistry **65**(10): 1429–1438.

Oh, K. H., S. H. Nam and H. S. Kim (2002). Directed evolution of N-carbamyl-
D-amino acid amidohydrolase for simultaneous improvement of oxidative
and thermal stability. *Biotechnol Prog* **18**(3): 413–417.

Ohrnberger, D. (1999). *The Bamboos of the World*. Amsterdam, the Netherlands:
Elsevier.

Okada, H., K. Tada, T. Sekiya, K. Yokoyama, A. Takahashi, H. Tohda, H. Kumagai and Y. Morikawa (1998). Molecular characterization and heterologous expression of the gene encoding a low-molecular-mass endoglucanase from *Trichoderma reesei* QM9414. *Appl Environ Microbiol* **64**(2): 555–563.

Oliveira, O. V., L. C. Freitas, T. P. Straatsma and R. D. Lins (2009). Interaction between the CBM of Cel9A from *Thermobifida fusca* and cellulose fibers. *J Mol Recognit* **22**(1): 38–45.

Olsen, M., B. Iverson and G. Georgiou (2000). High-throughput screening of enzyme libraries. *Curr Opin Biotechnol* **11**(4): 331–337.

Ooshima, H., K. Aso, Y. Harano and T. Yamamoto (1984). Microwave treatment of cellulosic materials for their enzymatic hydrolysis. *Biotechnol Lett* **6**(5): 289–294.

Otten, L. G. and W. J. Quax (2005). Directed evolution: Selecting today's biocatalysts. *Biomol Eng* **22**(1–3): 1–9.

Ouyang, J., M. Yan, D. Kong and L. Xu (2006). A complete protein pattern of cellulose and hemicellulase genes in the filamentous fungus *Trichoderma reesei*. *Biotechnol J* **1**(11): 1266–1274.

Packer, M. S. and D. R. Liu (2015). Methods for the directed evolution of proteins. *Nat Rev Genet* **16**(7): 379–394.

Pagès, S., A. Bélaïch, J.-P. Bélaïch, E. Morag, R. Lamed, Y. Shoham and E. A. Bayer (1997). Species-specificity of the cohesin-dockerin interaction between *Clostridium thermocellum* and *Clostridium cellulolyticum*: Prediction of specificity determinants of the dockerin domain. *Proteins Struct Func Bioinfor* **29**(4): 517–527.

Pagès, S., A. Bélaïch, H.-P. Fierobe, C. Tardif, C. Gaudin and J.-P. Bélaïch (1999). Sequence analysis of scaffolding protein CipC and ORFXp, a new cohesin-containing protein in *Clostridium cellulolyticum*: Comparison of various cohesin domains and subcellular localization of ORFXp. *J Bacteriol* **181**(6): 1801–1810.

Pan, X., N. Gilkes, J. Kadla, K. Pye, S. Saka, D. Gregg, K. Ehara, D. Xie, D. Lam and J. Saddler (2006). Bioconversion of hybrid poplar to ethanol and co-products using an organosolv fractionation process: Optimization of process yields. *Biotechnol Bioeng* **94**(5): 851–861.

Pan, X., D. Xie, N. Gilkes, D. J. Gregg and J. N. Saddler (2005). Strategies to enhance the enzymatic hydrolysis of pretreated softwood with high residual lignin content. *Appl Biochem Biotechnol* **121– 124**: 1069–1079.

Pan, X., D. Xie, R. W. Yu and J. N. Saddler (2008). The bioconversion of mountain pine beetle-killed lodgepole pine to fuel ethanol using the organosolv process. *Biotechnol Bioeng* **101**(1): 39–48.

Panneerselvam, A., R. R. Sharma-Shivappa, P. Kolar, D. A. Clare and T. Ranney (2013a). Hydrolysis of ozone pretreated energy grasses for optimal fermentable sugar production. *Bioresour Technol* **148**: 97–104.

Panneerselvam, A., R. R. Sharma-Shivappa, P. Kolar, T. Ranney and S. Peretti (2013b). Potential of ozonolysis as a pretreatment for energy grasses. *Bioresour Technol* **148**: 242–248.

Parsiegla, G., M. Juy, C. Reverbel-Leroy, C. Tardif, J. P. Belaich, H. Driguez and R. Haser (1998). The crystal structure of the processive endocellulase CelF of *Clostridium cellulolyticum* in complex with a thiooligosaccharide inhibitor at 2.0 Å resolution. *EMBO J* **17**(19): 5551–5562.

Pastell, H., L. Virkki, E. Harju, P. Tuomainen and M. Tenkanen (2009). Presence of 1-->3-linked 2-O-beta-d-xylopyranosyl-alpha-l-arabinofuranosyl side chains in cereal arabinoxylans. *Carbohydr Res* **344**(18): 2480–2488.

Paszczyński, A., V.-B. Huynh and R. Crawford (1986). Comparison of ligninase-I and peroxidase-M2 from the white-rot fungus *Phanerochaete chrysosporium*. *Arch Biochem Biophys* **244**(2): 750–765.

Patnaik, R., S. Louie, V. Gavrilovic, K. Perry, W. P. Stemmer, C. M. Ryan and S. del Cardayre (2002). Genome shuffling of *Lactobacillus* for improved acid tolerance. *Nat Biotechnol* **20**(7): 707–712.

Patton, J. (2010). Value-added coproducts from the production ofcellulosic ethanol, NDSU, Central Grasslands REC.

Paulechka, Y. U., G. J. Kabo, A. V. Blokhin, O. A. Vydrov, J. W. Magee and M. Frenkel (2003). Thermodynamic properties of 1-Butyl-3-methylimidazolium hexafluorophosphate in the ideal gas state. *J Chem Eng Data* **48**(3): 457–462.

Paulová, L., P. Patáková, M. Rychtera and K. Melzoch (2013). Production of 2nd generation of liquid biofuels. In *Liquid, Gaseous and Solid Biofuels - Conversion Techniques*, Z. Fang (Ed.). Rijeka, Croatia: InTech.

Pavelka, A., E. Chovancova and J. Damborsky (2009). HotSpot wizard: A web server for identification of hot spots in protein engineering. *Nucleic Acids Res* **37**(Web Server issue): W376–383.

Pearl, I. A. (1967). *The Chemistry of Lignin*. New York: Marcel Dekker.

Pedersen, M., K. S. Johansen and A. S. Meyer (2011). Low temperature lignocellulose pretreatment: Effects and interactions of pretreatment pH are critical for maximizing enzymatic monosaccharide yields from wheat straw. *Biotechnol Biofuels* **4**(11): 3–10.

Pedersen, M., A. Viksø-Nielsen and A. S. Meyer (2010). Monosaccharide yields and lignin removal from wheat straw in response to catalyst type and pH during mild thermal pretreatment. *Process Biochem* **45**(7): 1181–1186.

Pedrolli, D. B., A. C. Monteiro, E. Gomes and E. C. Carmona (2009). Pectin and pectinases: Production, characterization and industrial application of microbial pectinolytic enzymes. *Open Biotechnol J* **3**: 9–18.

Pei, X. Q., Z. L. Yi, C. G. Tang and Z. L. Wu (2011). Three amino acid changes contribute markedly to the thermostability of beta-glucosidase BglC from *Thermobifida fusca*. *Bioresour Technol* **102**(3): 3337–3342.

Peiji, G., Q. Yinbo, Z. Xin, Z. Mingtian and D. Yongcheng (1997). Screening microbial strain for improving the nutritional value of wheat and corn straws as animal feed. *Enzyme Microb Technol* **20**: 581–584.

Pell, G., M. P. Williamson, C. Walters, H. Du, H. J. Gilbert and D. N. Bolam (2003). Importance of hydrophobic and polar residues in ligand binding in the family 15 carbohydrate-binding module from *Cellvibrio japonicus* Xyn10C. *Biochemistry* **42**(31): 9316–9323.

Peng, H., Y. Zheng, M. Chen, Y. Wang, Y. Xiao and Y. Gao (2014). A starch-binding domain identified in α-amylase (AmyP) represents a new family of carbohydrate-binding modules that contribute to enzymatic hydrolysis of soluble starch. *FEBS Lett* **588**(7): 1161–1167.

Peng, P., F. Peng, J. Bian, F. Xu and R. Sun (2011). Studies on the starch and hemicelluloses fractionated by graded ethanol precipitation from bamboo *Phyllostachys bambusoides* f. shouzhu Yi. *J Agric Food Chem* **59**(6): 2680–2688.

Penttilä, M., P. Lehtovaara, H. Nevalainen, R. Bhikhabhai and J. Knowles (1986). Homology between cellulase genes of *Trichoderma reesei*: Complete nucleotide sequence of the endoglucanase I gene. *Gene* **45**(3): 253–263.

Perez, J. and T. W. Jeffries (1992). Roles of manganese and organic acid chelators in regulating lignin degradation and biosynthesis of peroxidases by *Phanerochaete chrysosporium*. *Appl Environ Microbiol* **58**(8): 2402–2409.

Pérez-Boada, M., F. J. Ruiz-Duenas, R. Pogni, R. Basosi, T. Choinowski, M. J. Martínez, K. Piontek and A. T. Martínez (2005). Versatile peroxidase oxidation of high redox potential aromatic compounds: Site-directed mutagenesis, spectroscopic and crystallographic investigation of three long-range electron transfer pathways. *J Mol Biol* **354**(2): 385–402.

Perrin, R. K. (2008). Ethanol and food prices—Preliminary assessment. In *Faculty Publications: Agricultural Economics*, Vol. 49, pp. 1–9. Lincoln, NE: University of Nebraska.

Perrone, P., C. M. Hewage, A. R. Thomson, K. Bailey, I. H. Sadler and S. C. Fry (2002). Patterns of methyl and O-acetyl esterification in spinach pectins: New complexity. *Phytochemistry* **60**: 67–77.

Peters, M. W., P. Meinhold, A. Glieder and F. H. Arnold (2003). Regio- and enantioselective alkane hydroxylation with engineered cytochromes P450 BM-3. *J Am Chem Soc* **125**(44): 13442–13450.

Petersen, B. O., S. Meier, J. Ø. Duus and M. H. Clausen (2008). Structural characterization of homogalacturonan by NMR spectroscopy—assignment of reference compounds. *Carbohydr Res* **343**(16): 2830–2833.

Peterson, P. B. (1988). Separation and characterization of botanical components of straw. *Agric Prog* **63**: 8–23.

Petre, J., R. Longin and J. Millet (1981). Purification and properties of an endo-β-1,4-glucanase from *Clostridium thermocellum*. *Biochimie* **63**(7): 629–639.

Petre, D., J. Millet, R. Longin, P. Béguin, H. Girard and J.-P. Aubert (1986). Purification and properties of the endoglucanase C of *Clostridium thermocellum* produced in *Escherichia coli*. *Biochimie* **68**(5): 687–695.

Phatak, L., K. C. Chang and G. Brown (1988). Isolation and characterization of pectin in sugar-beet pulp. *J Food Sci* **53**(3): 830–833.

Phillips, C. M., W. T. Beeson, J. H. Cate and M. A. Marletta (2011). Cellobiose dehydrogenase and a copper-dependent polysaccharide monooxygenase potentiate cellulose degradation by *Neurospora crassa*. *ACS Chem Biol* **6**(12): 1399–1406.

Pickersgill, R., J. Jenkins, G. Harris, W. Nasser and J. Robert-Baudouy (1994). The structure of *Bacillus subtilis* pectate lyase in complex with calcium. *Nat Struct Biol* **1**(10): 717–723.

Piggot, R. (2003). Treatment and fermentation of molasses when making rum-type spirits. In *The Alcohol Textbook*, K. A. Jacques, T. P. Lyons and D. R. Kelsall (Eds.). Bath: Nottingham University Press.

Pingali, S. V., V. S. Urban, W. T. Heller, J. McGaughey, H. O'Neill, M. Foston, D. A. Myles, A. Ragauskas and B. R. Evans (2010). Breakdown of cell wall nanostructure in dilute acid pretreated biomass. *Biomacromolecules* **11**(9): 2329–2335.

Pinheiro, B. A., H. J. Gilbert, K. Sakka, K. Sakka, V. O. Fernandes, J. A. Prates, V. D. Alves, D. N. Bolam, L. M. Ferreira and C. M. Fontes (2009). Functional insights into the role of novel type I cohesin and dockerin domains from *Clostridium thermocellum*. *Biochem J* **424**(3): 375–384.

Plechkova, N. V. and K. R. Seddon (2006). Ionic liquids: Designer solvents for green chemestry: An introduction. In *Methods and Reagents for Green Chemistry: An Introduction*, P. Tundo, A. Perosa and F. Zecchini (Eds.), pp. 105–130. New York: Wiley.

Pleiss, J. (2012). Rational design of enzymes. In *Enzyme Catalysis in Organic Synthesis*, K. Drauz, H. Gröger and O. May (Eds.), pp. 89–117. Weinheim, Germany: Wiley-VCH.

Pokhrel, S., J. C. Joo and Y. J. Yoo (2013). Shifting the optimum pH of *Bacillus circulans* xylanase towards acidic side by introducing arginine. *Biotechnol Bioprocess Eng* **18**(1): 35–42.

Pokkuluri, P. R., N. E. C. Duke, S. J. Wood, M. A. Cotta, X.-L. Li, P. Biely and M. Schiffer (2011). Structure of the catalytic domain of glucuronoyl esterase Cip2 from *Hypocrea jecorina*. *Proteins Struct Funct Bioinf* **79**(8): 2588–2592.

Polle, A., D. Janz, T. Teichmann and V. Lipka (2013). Poplar genetic engineering-promoting desirable wood characteristics and pest resistance. *Appl Microbiol Biotechnol* **97**(13): 5669–5679.

Popa, V. I. and J. Spiridon (1998). Hemicelluloses: Structure and properties. In *Polysaccharides: Structural Diversity and Functional Versatility*, S. Dumitriu (Ed.), pp. 297–311. New York: Marcel Dekker.

Popp, J. L. and T. K. Kirk (1991). Oxidation of methoxybenzenes by manganese peroxidase and by Mn^{3+}. *Arch Biochem Biophys* **288**(1): 145–148.

Popper, Z. A. and S. C. Fry (2005). Widespread occurrence of a covalent linkage between xyloglucan and acidic polysaccharides in suspension-cultured angiosperm cells. *Ann Bot* **96**(1): 91–99.

Popper, Z. A. and S. C. Fry (2008). Xyloglucan-pectin linkages are formed intra-protoplasmically, contribute to wall-assembly, and remain stable in the cell wall. *Planta* **227**(4): 781–794.

Pourbafrani, M., G. Forgacs, I. S. Horvath, C. Niklasson and M. J. Taherzadeh (2010). Production of biofuels, limonene and pectin from citrus wastes. *Bioresour Technol* **101**(11): 4246–4250.

Poutanen, K. (1988). An alpha-L-arabinofuranosidase of *Trichoderma reesei*. *J Biotechnol* **7**: 271–282.

Powell, D. A., E. R. Morris, M. J. Gidley and D. A. Rees (1982). Conformations and interactions of pectins: II. Influence of residue sequence on chain association in calcium pectate gels. *J Mol Biol* **155**(4): 517–531.

Prade, R. A., D. Zhan, P. Ayoubi and A. J. Mort (1999). Pectins, pectinases and plant-microbe interactions. *Biotechnol Genet Eng Rev* **16**: 361–391.

Prajapati, V. D., G. K. Jani, N. G. Moradiya, N. P. Randeria, B. J. Nagar, N. N. Naikwadi and B. C. Variya (2013). Galactomannan: A versatile biodegradable seed polysaccharide. *Int J Biol Macromol* **60**: 83–92.

Puri, V. P. (1984). Effect of crystallinity and degree of polymerization of cellulose on enzymatic saccharification. *Biotechnol Bioeng* **26**: 1219–1222.

Puri, V. P. and H. Mamers (1983). Explosive pretreatment of lignocellulosic residues with high-pressure carbon dioxide for the production of fermentation substrates. *Biotechnol Bioeng* **25**(12): 3149–3161.

Qi, M., H. S. Jun and C. W. Forsberg (2007). Characterization and synergistic interactions of *Fibrobacter succinogenes* glycoside hydrolases. *Appl Environ Microbiol* **73**(19): 6098–6105.

Qin, L., Z.-H. Liu, M. Jin, B.-Z. Li and Y.-J. Yuan (2013). High temperature aqueous ammonia pretreatment and post-washing enhance the high solids enzymatic hydrolysis of corn stover. *Bioresour Technol* **146**: 504–511.

Qing, Q. and C. E. Wyman (2011). Supplementation with xylanase and beta-xylosidase to reduce xylo-oligomer and xylan inhibition of enzymatic hydrolysis of cellulose and pretreated corn stover. *Biotechnol Biofuels* **4**(1): 18.

Quesada-Medina, J., F. J. López-Cremades and P. Olivares-Carrillo (2010). Organosolv extraction of lignin from hydrolyzed almond shells and application of the δ-value theory. *Bioresour Technol* **101**(21): 8252–8260.

Quin, M. B. and C. Schmidt-Dannert (2011). Engineering of biocatalysts—From evolution to creation. *ACS Catal* **1**(9): 1017–1021.

Quinlan, R. J., M. D. Sweeney, L. Lo Leggio, H. Otten, J. C. Poulsen, K. S. Johansen, K. B. Krogh et al. (2011). Insights into the oxidative degradation of cellulose by a copper metalloenzyme that exploits biomass components. *Proc Natl Acad Sci USA* **108**(37): 15079–15084.

Rabinovich, M. L., V. V. Nguen and A. A. Klesov (1982). Adsorption of cellulolytic enzymes on cellulose and the kinetics of the adsorbed enzymes. Two modes for interaction of the enzymes with the insoluble substrate. *Biokhimiia* **47**(3): 465–477.

Raíces, M., R. Montesino, J. Cremata, B. García, W. Perdomo, I. Szabó, G. Henriksson, B. M. Hallberg, G. Pettersson and G. Johansson (2002). Cellobiose quinone oxidoreductase from the white rot fungus *Phanerochaete chrysosporium* is produced by intracellular proteolysis of cellobiose dehydrogenase. *Biochim Biophys Acta Gene Struct Expr* **1576**(1–2): 15–22.

Raillard, S., A. Krebber, Y. Chen, J. E. Ness, E. Bermudez, R. Trinidad, R. Fullem et al. (2001). Novel enzyme activities and functional plasticity revealed by recombining highly homologous enzymes. *Chem Biol* **8**(9): 891–898.

Rajendran, K. and M. J. Taherzadeh (2014). Pretreatement of lignocellulosic materials. In *Bioprocessing of Renewable Resources to Commodity Bioproducts*, V. S. Bisaria and A. Kondo (Eds.). Hoboken, NJ: Wiley.

Ralet, M. C., J. F. Thibault, C. B. Faulds and G. Williamson (1994). Isolation and purification of feruloylated oligosaccharides from cell walls of sugar-beet pulp. *Carbohydr Res* **263**(2): 227–241.

Ralph, J., K. Lundquist, G. Brunow, F. Lu, H. Kim, P. Schatz, J. Marita et al. (2004). Lignins: Natural polymers from oxidative coupling of 4-hydroxyphenyl-propanoids. *Phytochem Rev* **3**(1–2): 29–60.

Ralph, J., J. Peng, F. Lu, R. D. Hatfield and R. F. Helm (1999). Are lignins optically active? *J Agric Food Chem* **47**(8): 2991–2996.

Raman, B., C. K. McKeown, M. Rodriguez, Jr., S. D. Brown and J. R. Mielenz (2011). Transcriptomic analysis of *Clostridium thermocellum* ATCC 27405 cellulose fermentation. *BMC Microbiol* **11**: 134.

Raman, B., C. Pan, G. B. Hurst, M. Rodriguez, Jr., C. K. McKeown, P. K. Lankford, N. F. Samatova and J. R. Mielenz (2009). Impact of pretreated switchgrass and biomass carbohydrates on *Clostridium thermocellum* ATCC 27405 cellulosome composition: A quantitative proteomic analysis. *PLoS One* **4**(4): e5271.

Ramirez, R. S., M. Holtzapple and N. Piamonte (2013). Fundamentals of biomass pretreatment at high pH. In *Aqueous Pretreatment of Plant Biomass for Biological and Chemical Conversion to Fuels and Chemicals*, C. E. Wyman (Ed.), pp. 169–200. New York: Wiley.

Ramos, L. P. (2003). The chemistry involved in the steam treatments of lignocellulosic materials. *Quim Nova* **26**(6): 863–871.

Ramos, L. P., C. Breuil and J. N. Saddler (1992). Comparison of steam pretreatment of eucalyptus, aspen, and spruce wood chips and their enzymatic hydrolysis. *Appl Biochem Biotechnol* **34/35**: 37–48.

Ramos, L. P. and J. N. Saddler (1994). Enzyme recycling during fed-batch hydrolysis of cellulose derived from steam-exploded Eucalyptus virninalls. *Appl Biochem Biotechnol* **45/ 46**: 193–207.

Ranney, J. W., L. L. Wright and P. A. Layton (1987). Hardwood energy crops: The technology of intensive culture. *J For* **85**(9): 17–28.

Rasmussen, L. E., H. R. Sørensen, J. Vind and A. Viksø-Nielsen (2006). Mode of action and properties of the beta-xylosidases from *Talaromyces emersonii* and *Trichoderma reesei*. *Biotechnol Bioeng* **94**(5): 869–876.

Reddy, G. V. B., M. Sridhar and M. H. Gold (2003). Cleavage of nonphenolic β-1 diarylpropane lignin model dimers by manganese peroxidase from *Phanerochaete chrysosporium*. *Eur J Biochem* **270**(2): 284–292.

Reese, E. T., R. G. Siu and H. S. Levinson (1950). The biological degradation of soluble cellulose derivatives and its relationship to the mechanism of cellulose hydrolysis. *J Bacteriol* **59**(4): 485–497.

Reetz, M. T., S. Prasad, J. D. Carballeira, Y. Gumulya and M. Bocola (2010a). Iterative saturation mutagenesis accelerates laboratory evolution of enzyme stereoselectivity: Rigorous comparison with traditional methods. *J Am Chem Soc* **132**(26): 9144–9152.

Reetz, M. T., P. Soni, L. Fernández, Y. Gumulya and J. D. Carballeira (2010b). Increasing the stability of an enzyme toward hostile organic solvents by directed evolution based on iterative saturation mutagenesis using the B-FIT method. *Chem Commun* **46**(45): 8657–8658.

Reetz, M. T., A. Zonta, K. Schimossek, K.-E. Jaeger and K. Liebeton (1997). Creation of enantioselective biocatalysts for organic chemistry by in vitro evolution. *Angew Chem Int Ed (English)* **36**(24): 2830–2832.

Reijnders, L. and M. A. J. Huijbregts (2009). *Biofuels for Road Transport: A Seed to Wheel Perspective*. London, UK: Springer-Verlag.

Reinhammar, B. R. M. (1972). Oxidation-reduction potentials of the electron acceptors in laccases and stellacyanin. *Biochim Biophys Acta Bioenergetics* **275**(2): 245–259.

Renewable Fuels Association (2014). *Falling Walls and Rising Tides. 2014 Ethanol Industry Outlook*. Washington, DC: Renewable Fuels Association.

Renganathan, V., K. Miki and M. H. Gold (1985). Multiple molecular forms of diarylpropane oxygenase, an H_2O_2-requiring, lignin-degrading enzyme from *Phanerochaete chrysosporium*. *Arch Biochem Biophys* **241**(1): 304–314.

Renganathan, V., K. Miki and M. H. Gold (1986). Role of molecular oxygen in lignin peroxidase reactions. *Arch Biochem Biophys* **246**(1): 155–161.

Reverbel-Leroy, C., S. Pages, A. Belaich, J. P. Belaich and C. Tardif (1997). The processive endocellulase CelF, a major component of the *Clostridium cellulolyticum* cellulosome: Purification and characterization of the recombinant form. *J Bacteriol* **179**(1): 46–52.

Richardson, J., J. E. K. Cooke, J. G. Isebrands, B. R. Thomas and K. C. J. Van Rees (2007). Poplar research in Canada—A historical perspective with a view to the future. *Can J Bot* **85**(12): 1136–1146.

Rignall, T. R., J. O. Baker, S. L. McCarter, W. S. Adney, T. B. Vinzant, S. R. Decker and M. E. Himmel (2002). Effect of single active-site cleft mutation on product specificity in a thermostable bacterial cellulase. In *Biotechnology for Fuels and Chemicals*, M. Finkelstein, J. McMillan and B. Davison (Eds.), pp. 383–394. Totowa, NJ: Humana Press.

Rivers, D. B. and G. H. Emert (1987). Lignocellulose pretreatment: A comparison of wet and dry ball attrition. *Biotechnol Lett* 9(5): 365–368.

Roberto, I. C., S. I. Mussatto and R. C. L. B. Rodrigues (2003). Dilute-acid hydrolysis for optimization of xylose recovery from rice straw in a semi-pilot reactor. *Ind Crops Prod* 17(3): 171–176.

Rodionova, N. A., I. M. Tavobilov and A. M. Bezborodov (1983). Beta-Xylosidase from *Aspergillus niger* 15: Purification and properties. *J Appl Biochem* 5(4–5): 300–312.

Rogers, R. D. and K. R. Seddon (2003). *Ionic Liquids as Green Solvents. Progress and Prospects*, ACS Symposium Series. Washington, DC: American Chemical Society.

Rogowski, A., A. Baslé, C. S. Farinas, A. Solovyova, J. C. Mortimer, P. Dupree, H. J. Gilbert and D. N. Bolam (2014). Evidence that GH115 α-glucuronidase activity, which is required to degrade plant biomass, is dependent on conformational flexibility. *J Biol Chem* 289(1): 53–64.

Roldán, A., V. Palacios, X. Peñate, T. Benítez and L. Pérez (2009). Use of *Trichoderma enzymatic* extracts on vinification of *Palomino fino* grapes in the sherry region. *J Food Eng* 75(3): 375–382.

Rollin, J. A., Z. Zhu, N. Sathitsuksanoh and Y. H. Zhang (2011). Increasing cellulose accessibility is more important than removing lignin: A comparison of cellulose solvent-based lignocellulose fractionation and soaking in aqueous ammonia. *Biotechnol Bioeng* 108(1): 22–30.

Romaniec, M. P., U. Fauth, T. Kobayashi, N. S. Huskisson, P. J. Barker and A. L. Demain (1992). Purification and characterization of a new endoglucanase from *Clostridium thermocellum*. *Biochem J* 283(Pt 1): 69–73.

Roncero, M. B., A. L. Torres, J. F. Colom and T. Vidal (2005). The effect of xylanase on lignocellulosic components during the bleaching of wood pulps. *Bioresour Technol* 96(1): 21–30.

Rorick, R., N. Nahar and S. W. Pryor (2011). Ethanol production from sugar beet pulp using Escherichiacoli KO11 and Saccharomy cescerevisiae. *Biol Eng Trans* 3(4): 199–209.

Rouvinen, J., T. Bergfors, T. Teeri, J. K. Knowles and T. A. Jones (1990). Three-dimensional structure of cellobiohydrolase II from *Trichoderma reesei*. *Science* 249(4967): 380–386.

Rouyi, C., S. Baiya, S. K. Lee, B. Mahong, J. S. Jeon, J. R. Ketudat-Cairns and M. Ketudat-Cairns (2014). Recombinant expression and characterization of the cytoplasmic rice β-glucosidase Os1BGlu4. *PLoS One* 9(5): e96712.

Rowell, R. M., R. Petersen and M. A. Tshabalala (2012). Cell wall chemistry. In *Handbook of Wood Chemistry and Wood Composites*, R. M. Rowell (Ed.), pp. 33–72. Boca Raton, FL: CRC Press, Taylor & Francis Group.

Ruelius, H. W., R. M. Kerwin and F. W. Janssen (1968). Carbohydrate oxidase, a novel enzyme from polyporus obtusus: I. Isolation and purification. *Biochim Biophys Acta* 167(3): 493–500.

Rughani, J. and G. D. McGinnis (1989). Combined rapid-steam hydrolysis and organosolv pretreatment of mixed southern hardwoods. *Biotechnol Bioeng* 33(6): 681–686.

Ruiz-Duenas, F. J., M. J. Martinez and A. T. Martinez (1999). Molecular characterization of a novel peroxidase isolated from the ligninolytic fungus *Pleurotus
eryngii*. *Mol Microbiol* **31**(1): 223–235.

Ruiz-Duenas, F. J., M. Morales, E. Garcia, Y. Miki, M. J. Martinez and A. T. Martinez
(2009). Substrate oxidation sites in versatile peroxidase and other basidiomycete peroxidases. *J Exp Bot* **60**(2): 441–452.

Ruller, R., J. Alponti, L. A. Deliberto, L. M. Zanphorlin, C. B. Machado and R. J.
Ward (2014). Concommitant adaptation of a GH11 xylanase by directed evolution to create an alkali-tolerant/thermophilic enzyme. *Protein Eng Des Sel*
27(8): 255–262.

Runge, C. F. and B. Senauer (2007). How biofuels could starve the poor. *Foreign
Affairs* **86**: 41–53.

Ruscio, J. Z., J. E. Kohn, K. A. Ball and T. Head-Gordon (2009). The influence of protein dynamics on the success of computational enzyme design. *J Am Chem
Soc* **131**(39): 14111–14115.

Ryabova, O., M. Vršanská, S. Kaneko, W. H. van Zyl and P. Biely (2009). A novel
family of hemicellulolytic α-glucuronidase. *FEBS Lett* **583**(9): 1457–1462.

Saha, B. C. (2003a). Hemicellulose bioconversion. *J Ind Microbiol Biotechnol* **30**(5):
279–291.

Saha, B. C. (2003b). Purification and properties of an extracellular beta-xylosidase
from a newly isolated Fusarium proliferatum. *Bioresour Technol* **90**(1): 33–98.

Saha, B. C. and R. J. Bothast (1998). Purification and characterization of a novel
thermostable α-L-arabinofuranosidase from a color-variant strain of
Aureobasidium pullulans. *Appl Environ Microbiol* **64**(1): 216–220.

Saha, B. C. and M. A. Cotta (2006). Ethanol production from alkaline peroxide pretreated enzymatically saccharified wheat straw. *Biotechnol Prog* **22**(2): 449–453.

Saharay, M., H. Guo and J. C. Smith (2010). Catalytic mechanism of cellulose degradation by a cellobiohydrolase, CelS. *PLoS One* **5**(10): e12947.

Sakamoto, O. (2004). *The financial feasibility analysis of municipal solid waste to ethanol
conversion*. Master of science thesis, Michigan State University.

Sakon, J., D. Irwin, D. B. Wilson and P. A. Karplus (1997). Structure and mechanism of endo/exocellulase E4 from *Thermomonospora fusca*. *Nat Struct Biol*
4(10): 810–818.

Sala, O. E., D. Sax and H. Leslie (2009). Biodiversity consequences of increased biofuel production. In *Biofuels: Environmental Consequences and Interactions with
Changing Land Use*, R. H. S. Bringezu (Ed.), 127–137. Ithaca: Scope.

Salamitou, S., M. Lemaire, T. Fujino, H. Ohayon, P. Gounon, P. Beguin and J. P.
Aubert (1994). Subcellular localization of *Clostridium thermocellum* ORF3p, a
protein carrying a receptor for the docking sequence borne by the catalytic
components of the cellulosome. *J Bacteriol* **176**(10): 2828–2834.

Salamitou, S., K. Tokatlidis, P. Béguin and J.-P. Aubert (1992). Involvement of separate domains of the cellulosomal protein S1 of *Clostridium thermocellum* in
binding to cellulose and in anchoring of catalytic subunits to the cellulosome. *FEBS Lett* **304**(1): 89–92.

Saloheimo, A., B. Henrissat, A.-M. Hoffrén, O. Teleman and M. Penttilä (1994).
A novel, small endoglucanase gene, egl5, from *Trichoderma reesei* isolated by
expression in yeast. *Mol Microbiol* **13**(2): 219–228.

Saloheimo, M., J. Kuja-Panula, E. Ylösmäki, M. Ward and M. Penttilä (2002).
Enzymatic properties and intracellular localization of the novel *Trichoderma
reesei* β-glucosidase BGLII (Cel1A). *Appl Environ Microbiol* **68**(9): 4546–4553.

Saloheimo, M., P. Lehtovaara, M. Penttilä, T. T. Teeri, J. Ståhlberg, G. Johansson, G. Pettersson, M. Claeyssens, P. Tomme and J. K. C. Knowles (1988). EGIII, a new endoglucanase from *Trichoderma reesei*: The characterization of both gene and enzyme. *Gene* **63**(1): 11–21.

Saloheimo, M., T. Nakari-SetäLä, M. Tenkanen and M. Penttilä (1997). cDNA cloning of a *Trichoderma reesei* cellulase and demonstration of endoglucanase activity by expression in yeast. *Eur J Biochem* **249**(2): 584–591.

Salvachúa, D., A. Prieto, Á. T. Martínez and M. J. Martínez (2013). Characterization of a novel dye-decolorizing peroxidase (DyP)-type enzyme from *Irpex lacteus* and its application in enzymatic hydrolysis of wheat straw. *Appl Environ Microbiol* **79**(14): 4316–4324.

Salvador, A. C., M. d. C. Santos and J. A. Saraiva (2010). Effect of the ionic liquid [bmim]Cl and high pressure on the activity of cellulase. *Green Chem* **12**: 632–635.

Samejima, M. and K. E. Eriksson (1992). A comparison of the catalytic properties of cellobiose:quinone oxidoreductase and cellobiose oxidase from *Phanerochaete chrysosporium*. *Eur J Biochem* **207**(1): 103–107.

Sanchez, O. J. and C. A. Cardona (2008). Trends in biotechnological production of fuel ethanol from different feedstocks. *Bioresour Technol* **99**(13): 5270–5295.

Sannigrahi, P., S. J. Miller and A. J. Ragauskas (2010a). Effects of organosolv pretreatment and enzymatic hydrolysis on cellulose structure and crystallinity in Loblolly pine. *Carbohydr Res* **345**(7): 965–970.

Sannigrahi, P., A. J. Ragauskas and G. A. Tuskan (2010b). Poplar as a feedstock for biofuels: A review of compositional characteristics. *Biofuels Bioprod Biorefin* **4**(2): 209–226.

Santos, A., S. Mendes, V. Brissos and L. O. Martins (2014). New dye-decolorizing peroxidases from *Bacillus subtilis* and *Pseudomonas putida* MET94: Towards biotechnological applications. *Appl Microbiol Biotechnol* **98**(5): 2053–2065.

Santos, R. B., E. A. Capanema, M. Y. Balakshin, H.-m. Chang and H. Jameel (2012). Lignin structural variation in hardwood species. *J Agric Food Chem* **60**: 4923–4930.

Saritha, M. and A. A. Lata (2012). Biological pretreatment of lignocellulosic substrates for enhanced delignification and enzymatic digestibility. *Indian J Microbiol* **52**(2): 122–130.

Sathitsuksanoh, N., A. George and Y. H. P. Zhang (2013). New lignocellulose pretreatments using cellulose solvents: A review. *J Chem Technol Biotechnol* **88**(2): 169–180.

Sathitsuksanoh, N., Z. Zhu, T.-J. Ho, M.-D. Bai and Y.-H. P. Zhang (2010). Bamboo saccharification through cellulose solvent-based biomass pretreatment followed by enzymatic hydrolysis at ultra-low cellulase loadings. *Bioresour Technol* **101**(13): 4926–4929.

Sathitsuksanoh, N., Z. Zhu, N. Templeton, J. A. Rollin, S. P. Harvey and Y. H. P. Zhang (2009). Saccharification of a potential bioenergy crop, *Phragmites australis* (Common Reed), by lignocellulosefractionation followed by enzymatic hydrolysis at decreased cellulase loadings. *Ind Eng Chem Res* **48**(13): 6441–6447.

Sathitsuksanoh, N., Z. Zhu, S. Wi and Y. H. Zhang (2011). Cellulose solvent-based biomass pretreatment breaks highly ordered hydrogen bonds in cellulose fibers of switchgrass. *Biotechnol Bioeng* **108**(3): 521–529.

Sathitsuksanoh, N., Z. Zhu and Y. H. P. Zhang (2012). Cellulose solvent- and organic solvent-based lignocellulose fractionation enabled efficient sugar release from a variety of lignocellulosic feedstocks. *Bioresour Technol* **117**: 228–233.

Satoshi, O. (1992). *The Search for Bioactive Compounds from Microorganisms*. New York: Springer.

Satyanarayana, D. V. T. (2013). Improvement in thermostability of metagenomic GH11 endoxylanase (Mxyl) by site-directed mutagenesis and its applicability in paper pulp bleaching process. *J Ind Microbiol Biotechnol* **40**(12): 1373–1381.

Savile, C. K., J. M. Janey, E. C. Mundorff, J. C. Moore, S. Tam, W. R. Jarvis, J. C. Colbeck et al. (2010). Biocatalytic asymmetric synthesis of chiral amines from ketones applied to sitagliptin manufacture. *Science* **329**(5989): 305–309.

Sawada, T. and Y. Nakamura (2001). Low energy steam explosion treatment of plant biomass. *J Chem Technol Biotechnol* **76**(2): 139–146.

Scalbert, A., B. Monties, J.-Y. Lallemand, E. Guittet and C. Rolando (1985). Ether linkage between phenolic acids and lignin fractions from wheat straw. *Phytochemistry* **24**(6): 1359–1362.

Schaeffer, F., M. Matuschek, G. Guglielmi, I. Miras, P. M. Alzari and P. Béguin (2002). Duplicated dockerin subdomains of *Clostridium thermocellum* endoglucanase CelD bind to a cohesin domain of the scaffolding protein CipA with distinct thermodynamic parameters and a negative cooperativity. *Biochemistry* **41**(7): 2106–2114.

Schaffner, D. W. and R. T. Toledo (1991). Cellulase production by *Trichoderma reesei* when cultured on xylose-based media supplemented with sorbose. *Biotechnol Bioeng* **37**(1): 12–16.

Scheller, H. V. and P. Ulvskov (2010). Hemicelluloses. *Annu Rev Plant Biol* **61**(1): 263–289.

Schifreen, R. S., D. A. Hanna, L. D. Bowers and P. W. Carr (1977). Analytical aspects of immobilized enzyme columns. *Anal Chem* **49**(13): 1929–1939.

Schimming, S., W. H. Schwarz and W. L. Staudenbauer (1991). Properties of a thermoactive beta-1,3-1,4-glucanase (lichenase) from *Clostridium thermocellum* expressed in *Escherichia coli*. *Biochem Biophys Res Commun* **177**(1): 447–452.

Schiraldi, C. and M. De Rosa (2002). The production of biocatalysts and biomolecules from extremophiles. *Trends Biotechnol* **20**(12): 515–521.

Schmer, M. R., K. P. Vogel, R. B. Mitchell and R. K. Perrin (2008). Net energy of cellulosic ethanol from switchgrass. *Proc Natl Acad Sci USA* **105**(2): 464–469.

Schneider, P., M. B. Caspersen, K. Mondorf, T. Halkier, L. K. Skov, P. R. Østergaard, K. M. Brown, S. H. Brown and F. Xu (1999). Characterization of a *Coprinus cinereus* laccase. *Enzyme Microb Technol* **25**(6–25): 502–508.

Schoch, T. J. (1942). Fractionation of starch by selective precipitation with butanol. *J Am Chem Soc* **64**: 2957–2961.

Schoemaker, H. E., P. J. Harvey, R. M. Bowen and J. M. Palmer (1985). On the mechanism of enzymatic fignin breakdown *Febs Lett* **183**(1): 7–12.

Schols, H. A., E. J. Bakx, D. Schipper and A. G. J. Voragen (1995). A xylogalacturonan subunit present in the modified hairy regions of apple pectin. *Carbohydr Res* **279**: 265–279.

Schols, H. A., C. C. J. M. Geraeds, M. F. Searle-van Leeuwen, F. J. M. Kormelink and A. G. J. Voragen (1990). Rhamnogalacturonase: A novel enzyme that degrades the hairy regions of pectins. *Carbohydr Res* **206**(1): 105–115.

Schuerch, C. (1963). Plasticizing wood with liquid ammonia. *J Ind Eng Chem* **55**: 39.

Schulze, E. (1891). Information regarding chemical composition of plant cell membrane. *Ber Dtsch Chem Ges* **24**: 2277–2287.

Schurz, J. (1978). Bioconversion of cellulosic substances into energy chemicals and microbial protein. *Symposium Proceedings*, IIT, New Delhi, India.

Schuster, A. and M. Schmoll (2010). Biology and biotechnology of Trichoderma. *Appl Microbiol Biotechnol* **87**(3): 787–799.

Schwarz, W., K. Bronnenmeier and W. L. Staudenbauer (1985). Molecular cloning of *Clostridium thermocellum* genes involved in β-glucan degradation in bacteriophage lambda. *Biotechnol Lett* **7**(12): 859–864

Schwarz, W. H., F. Gräbnitz and W. L. Staudenbauer (1986). Properties of a *Clostridium thermocellum* endoglucanase produced in *Escherichia coli*. *Appl Environ Microbiol* **51**(6): 1293–1299.

Schwarz, W. H., S. Schimming, K. P. Rücknagel, S. Burgschwaiger, G. Kreil and W. L. Staudenbauer (1988). Nucleotide sequence of the celC gene encoding endoglucanase C of *Clostridium thermocellum*. *Gene* **63**(1): 23–30.

Scurlock, J. M. O., D. C. Dayton and B. Hames (2000). Bamboo: An overlooked biomass resource? *Biomass Bioenergy* **19**(4): 229–244.

Searle-van Leeuwen, M. J. F., L. A. M. Broek, H. A. Schols, G. Beldman and A. G. J. Voragen (1992). Rhamnogalacturonan acetylesterase: A novel enzyme from *Aspergillus aculeatus*, specific for the deacetylation of hairy (ramified) regions of pectins. *Appl Microbiol Biotechnol* **38**(3): 347–349.

Seegmiller, C. G. and E. F. Jansen (1952). Polymethylgalacturonase an enzyme causing the glycosidic hydrolysis of esterified pectic substances. *J Biol Chem* **195**(1): 327–333.

Seelig, B. (2011). mRNA display for the selection and evolution of enzymes from in vitro-translated protein libraries. *Nat Protoc* **6**(4): 540–552.

Seiboth, B., L. Hartl, N. Salovuori, K. Lanthaler, G. D. Robson, J. Vehmaanperä, M. E. Penttilä and C. P. Kubicek (2005). Role of the bga1-encoded extracellular β-galactosidase of *Hypocrea jecorina* in cellulase induction by lactose. *Appl Environ Microbiol* **71**(2): 851–857.

Seiboth, B., S. Herold and C. P. Kubicek (2012). Metabolic engineering of inducer formation for cellulase and hemicellulase gene expression in *Trichoderma reesei*. In *Subcellular Biochemistry. Reprogramming Microbial Metabolic Pathways*, X. Wang, J. Chen and P. Quinn (Eds.), Vol. 64, pp. 367–390. Dordrecht, the Netherlands: Springer Science+Business Media.

Selby, K. and C. C. Maitland (1967). The cellulase of *Trichoderma viride*. Separation of the components involved in the solubilization of cotton. *Biochem J* **104**: 716–724.

Selvendran, R. R. and P. Ryden (1990). Isolation and analysis of plant cell walls. In *Methods in Plant Biochemistry*, P. M. Dey (Ed.). London: Academic Press

Seyedarabi, A., T. T. To, S. Ali, S. Hussain, M. Fries, R. Madsen, M. H. Clausen, S. Teixteira, B. K. and R. W. Pickersgill (2010). Structural insights into substrate specificity and the anti beta-elimination mechanism of pectate lyase. *Biochemistry* **49**(3): 539–546.

Shafiei, M., R. Kumar and K. Karimi (2015). Pretreatment of lignocellulosic biomass. In *Lignocellulose-Based Bioproducts*, K. Karimi (Ed.), pp. 85–154. Cham, Switzerland: Springer International Publishing.

Shallom, D. and Y. Shoham (2003). Microbial hemicellulases. *Curr Opin Microbiol* **6**(3): 219–228.

Shamsudin, S., U. K. M. Shah, H. Zainudin, S. Abd-Aziz, S. M. M. Kamal, Y. Shirai and M. A. Hassan (2012). Effect of steam pretreatment on oil palm empty fruit bunch for the production of sugars. *Biomass Bioenergy* **36**: 280–288.

Shao, X., M. Jin, A. Guseva, C. Liu, V. Balan, D. Hogsett, B. E. Dale and L. Lynd (2011). Conversion for Avicel and AFEX pretreated corn stover by *Clostridium thermocellum* and simultaneous saccharification and fermentation: Insights into microbial conversion of pretreated cellulosic biomass. *Bioresour Technol* **102**(17): 8040–8045.

Shapouri, H., M. Salassi and N. J. (2006). *The Economic Feasibility of Ethanol Production from Sugar in the United States*. Washington, DC: U.S. Department of Agriculture (USDA).

Sharma, N., M. Rathore and M. Sharma (2013a). Microbial pectinase: Sources, characterization and applications. *Rev Environ Sci Bio/Technol* **12**(1): 45–60.

Sharma, R., V. Palled, R. R. Sharma-Shivappa and J. Osborne (2013b). Potential of potassium hydroxide pretreatment of switchgrass for fermentable sugar production. *Appl Biochem Biotechnol* **169**(3): 761–772.

Sheth, K. and J. K. Alexander (1967). Cellodextrin phosphorylase from *Clostridium thermocellum*. *Biochim Biophys Acta* **148**(3): 808–810.

Sheth, K. and J. K. Alexander (1969). Purification and properties of β-1,4-oligoglucan: Orthophosphate glucosyltransferase from *Clostridium thermocellum*. *J Biol Chem* **244**(2): 457–464.

Shevchik, V. E. and N. Hugouvieux-Cotte-Pattat (1997). Identification of a bacterial pectin acetyl esterase in *Erwinia chrysanthemi* 3937. *Mol Microbiol* **24**(6): 1285–1301.

Shi, A. Z., L. P. Koh and H. T. W. Tan (2009a). The biofuel potential of municipal solid waste. *GCB Bioenergy* **1**(5): 317–320.

Shi, H., H. Ding, Y. Huang, L. Wang, Y. Zhang, X. Li and F. Wang (2014). Expression and characterization of a GH43 endo-arabinanase from Thermotoga thermarum. *BMC Biotechnol* **14**: 35.

Shi, J., M. Ebrik, B. Yang and C. E. Wyman (2009b). *The Potential of Cellulosic Ethanol Production from Municipal Solid Waste: A Technical and Economic Evaluation*. Berkeley, CA: University of California Energy Institute.

Shimizu, K., M. Hashi and K. Sakurai (1978). Isolation from a softwood xylan of oligosaccharides containing two 4-O-methyl-d-glucuronic acid residues. *Carbohydr Res* **62**(1): 117–126.

Shin, S.-J. and Y. J. Sung (2010). Improving enzymatic saccharification of hybrid poplar by electron beam irradiation pretreatment. *J Biobased Mater Bioenergy* **4**(1): 23–26.

Shinmyo, A., D. V. Garcia-Martinez and A. L. Demain (1979). Studies on the extracellular cellulolytic enzyme complex produced by *Clostridium thermocellum*. *J Appl Biochem* **1**: 202–209.

Shoemaker, S., V. Schweickart, M. Ladner, D. Gelfand, S. Kwok, K. Myambo and M. Innis (1983). Molecular cloning of exo-cellobiohydrolase I derived from *Trichoderma reesei* strain L27. *Nat Biotech* **1**(8): 691–696.

Shoham, Y., R. Lamed and E. A. Bayer (1999). The cellulosome concept as an efficient microbial strategy for the degradation of insoluble polysaccharides. *Trends Microbiol* **7**(7): 275–281.

Shoseyov, O., M. Takagi, M. A. Goldstein and R. H. Doi (1992). Primary sequence analysis of *Clostridium cellulovorans* cellulose binding protein A. *Proc Natl Acad Sci USA* **89**(8): 3483–3487.

Shuai, L., Q. Yang, J. Y. Zhu, F. C. Lu, P. J. Weimer, J. Ralph and X. J. Pan (2010). Comparative study of SPORL and dilute-acid pretreatments of spruce for cellulosic ethanol production. *Bioresour Technol* **101**(9): 3106–3114.

Siau, J. W., S. Chee, H. Makhija, C. M. Wai, S. H. Chandra, S. Peter, P. Dröge and F. J. Ghadessy (2015). Directed evolution of λ integrase activity and specificity by genetic derepression. *Protein Eng Des Sel* **28**(7): 211–220.

Siddhartha Bhatt, M. and N. Rajkumar (2001). Mapping of combined heat and power systems in cane sugar industry. *Appl Therm Eng* **21**(17): 1707–1719.

Sieber, V., C. A. Martinez and F. H. Arnold (2001). Libraries of hybrid proteins from distantly related sequences. *Nat Biotechnol* **19**(5): 456–460.

Sih, C. J. and R. H. McBee (1955). A cellobiose-phosphorylase in *Clostridium thermocellum*. *Proc Montana Acad Sci* **15**: 21–22.

Silverstein, R. A., Y. Chen, R. R. Sharma-Shivappa, M. D. Boyette and J. Osborne (2007). A comparison of chemical pretreatment methods for improving saccharification of cotton stalks. *Bioresour Technol* **98**(16): 3000–3011.

Simpson, P. J., H. Xie, D. N. Bolam, H. J. Gilbert and M. P. Williamson (2000). The structural basis for the ligand specificity of family 2 carbohydrate-binding modules. *J Biol Chem* **275**(52): 41137–41142.

Sims, I. M., D. J. Craik and A. Bacic (1997). Structural characterisation of galacto-glucomannan secreted by suspension-cultured cells of *Nicotiana plumbaginifolia*. *Carbohydr Res* **303**(1): 79–92.

Sims, I. M., S. L. A. Munro, G. Currie, D. Craik and A. Bacic (1996). Structural characterisation of xyloglucan secreted by suspension-cultured cells of *Nicotiana plumbaginifolia*. *Carbohydr Res* **293**(2): 147–172.

Singh, J. S., W. K. Lauenroth and D. G. Milchunas (1983). Geography of grassland ecosystems. *Prog Phys Geogr* **7**(1): 46–80.

Singh, N. and M. Cheryan (1998). Extraction of oil from corn distillers dried grains with soluble. *Trans ASAE* **41**(6): 1775–1777.

Singh, R. N. and V. K. Akimenko (1993). Isolation of a cellobiohydrolase of *Clostridium thermocellum* capable of degrading natural crystalline substrates. *Biochem Biophys Res Commun* **192**(3): 1123–1130.

Singh, R. N. and V. K. Akimenko (1994). Isolation and characterization of a complex forming hydrophilic endoglucanase of *Clostridium thermocellum*. *Biochem Mol Biol Int* **32**(3): 409–417.

Singh, S., B. A. Simmons and K. P. Vogel (2009). Visualization of biomass solubilization and cellulose regeneration during ionic liquid pretreatment of switchgrass. *Biotechnol Bioeng* **104**(1): 68–75.

Singh, S. K., C. Heng, J. D. Braker, V. J. Chan, C. C. Lee, D. B. Jordan, L. Yuan and K. Wagschal (2014). Directed evolution of GH43 β-xylosidase XylBH43 thermal stability and L186 saturation mutagenesis. *J Ind Microbiol Biotechnol* **41**(3): 489–498.

Singha, J., M. Suhag and A. Dhaka (2015). Augmented digestion of lignocellulose by steam explosion, acid and alkaline pretreatment methods: A review. *Carbohydr Polym* **117**: 624–631.

Sipos, B., Z. Benkő, D. Dienes, K. Réczey, L. Viikari and M. Siika-aho (2010). Characterisation of specific activities and hydrolytic properties of cell-wall-degrading enzymes produced by *Trichoderma reesei* Rut C30 on different carbon sources. *Appl Biochem Biotechnol* **161**(1–8): 347–364.

Sivers, M. V. and G. Zacchi (1995). A techno-economical comparison of three processes for the production of ethanol from pine. *Bioresour Technol* **51**(1): 43–52.

Sjöström, E. (1993). *Wood Chemistry—Fundamentals and Applications*. San Diego, CA: Academic Press.

Smiley, J. A. and S. J. Benkovic (1994). Selection of catalytic antibodies for a biosynthetic reaction from a combinatorial cDNA library by complementation of an auxotrophic *Escherichia coli*: Antibodies for orotate decarboxylation. *Proc Natl Acad Sci USA* **91**(18): 8319–8323.

Smith, J. E., P. D. Miles, C. H. Perry and S. A. Pugh (2009). Forest resources of the United States, 2007: A technical document supporting the forest service 2010 RPA Assessment. Washington, DC: General Technical Report-USDA Forest Service WO-78.

Smith, M. A., A. Rentmeister, C. D. Snow, T. Wu, M. F. Farrow, F. Mingardon and F. H. Arnold (2012). A diverse set of family 48 bacterial glycoside hydrolase cellulases created by structure-guided recombination. *FEBS J* **279**(24): 4453–4465.

Smith, R. (2013). Eastern Iowa developer plans to turn trash into ethanol, compressed natural gas. *The Gazette*.

Smith, S. P. and E. A. Bayer (2013). Insights into cellulosome assembly and dynamics: From dissection to reconstruction of the supramolecular enzyme complex. *Curr Opin Struct Biol* **23**(5): 686–694.

Snoek, T., M. Picca Nicolino, S. Van den Bremt, S. Mertens, V. Saels, A. Verplaetse, J. Steensels and K. J. Verstrepen (2015). Large-scale robot-assisted genome shuffling yields industrial *Saccharomyces cerevisiae* yeasts with increased ethanol tolerance. *Biotechnol Biofuels* **8**: 32.

Soderstrom, T. R. and C. E. Calderon (1979). Ecology and phytosociology of bamboo vegetation In *Ecology of Grasslands and Bamboo Lands in the World*, N. Numata (Ed.), pp. 223–236. Jena, Germany: VEB Gustav Ficher Verlag.

Sokhansanj, S., S. Mani, A. Turhollow, A. Kumar, D. Bransby, L. Lynd and M. Laser (2009). Large-scale production, harvest and logistics of switchgrass (*Panicum virgatum* L.)—Current technology and envisioning a mature technology. *Biofuels Bioprod Biorefin* **3**(2): 124–141.

Solbak, A. I., T. H. Richardson, R. T. McCann, K. A. Kline, F. Bartnek, G. Tomlinson, X. Tan et al. (2005). Discovery of pectin-degrading enzymes and directed evolution of a novel pectate lyase for processing cotton fabric. *J Biol Chem* **280**(10): 9431–9438.

Solms, J. and H. Deuel (1955). Über den mechanismus der enzymatischen verseifung von pektinstoffen. *Helv Chim Acta* **38**(1): 321–329.

Solomon, E. I., P. Chen, M. Metz, S. K. Lee and A. E. Palmer (2001). Oxygen binding, activation, and reduction to water by copper proteins. *Angew Chem Int Ed Engl* **40**(24): 4570–4590.

Solomon, E. I., U. M. Sundaram and T. E. Machonkin (1996). Multicopper oxidases and oxygenases. *Chem Rev* **96**(7): 2563–2605.

Song, L., S. Laguerre, C. Dumon, S. Bozonnet and M. J. O'Donohue (2010). A high-throughput screening system for the evaluation of biomass-hydrolyzing glycoside hydrolases. *Bioresour Technol* **101**(21): 8237–8243.

Soutschek-Bauer, E. and W. L. Staudenbauer (1987). Synthesis and secretion of a heat-stable carboxymethylcellulase from *Clostridium thermocellum* in *Bacillus subtilis* and *Bacillus stearothermophilus*. *Mol Gen Genet* **208**(3): 537–541.

Špániková, S. and P. Biely (2006). Glucuronoyl esterase—Novel carbohydrate esterase produced by Schizophyllum commune. *FEBS Lett* **580**(19): 4597–4601.

Spence, A. K., J. Boddu, D. Wang, B. James, K. Swaminathan, S. P. Moose and S. P. Long (2014). Transcriptional responses indicate maintenance of photosynthetic proteins as key to the exceptional chilling tolerance of C4 photosynthesis in *Miscanthus* × *giganteus*. *J Exp Bot* **65**(13): 3737–3747.

Spinelli, S., H. P. Fierobe, A. Belaich, J. P. Belaich, B. Henrissat and C. Cambillau (2000). Crystal structure of a cohesin module from *Clostridium cellulolyticum*: Implications for dockerin recognition. *J Mol Biol* **304**(2): 189–200.

Spinnler, H. E., B. Lavigne and H. Blachere (1986). Pectinolytic activity of *Clostridium thermocellum*: Its use for anaerobic fermentation of sugar beet pulp. *Appl Microbiol Biotechnol* **23**(6): 434–437.

Srebotnik, E., K. Messner and R. Foisner (1988). Penetrability of white rot-degraded pine wood by the lignin peroxidase of *Phanerochaete chrysosporium*. *Appl Environ Microbiol* **54**(11): 2608–2614.

Srinivasan, N. and L.-K. Ju (2010). Pretreatment of guayule biomass using super-critical carbon dioxide-based method. *Bioresour Technol* **101**: 9785–9791.

Srivastava, S., N. Ghosh and G. Pal (2013). Metagenomics: Mining environmental genomes. In *Biotechnology for Environmental Management and Resource Recovery*, R. C. Kuhad and A. Singh (Eds.), pp. 161–189. New Delhi, India: Springer.

Stahl, S. W., M. A. Nash, D. B. Fried, M. Slutzki, Y. Barak, E. A. Bayer and H. E. Gaub (2012). Single-molecule dissection of the high-affinity cohesin-dockerin complex. *Proc Natl Acad Sci USA* **109**(50): 20431–20436.

Stahlberg, J., G. Johansson and G. Pettersson (1988). A binding-site-deficient, cata-lytically active, core protein of endoglucanase III from the culture filtrate of *Trichoderma reesei Eur J Biochem* **173**: 179–183.

Stahlberg, J., G. Johansson and G. Pettersson (1993). *Trichoderma reesei* has no true exo-cellulase: All intact and truncated cellulases produce new reducing end groups on cellulose. *Biochim Biophys Acta* **1157**(1): 107–113.

Stålbrand, H. (1993). Purification and characterization of two β-mannanases from *Trichoderma reesei*. *J Biotechnol* **29**(3): 229–242.

Stålbrand, H., A. Saloheimo, J. Vehmaanperä, B. Henrissat and M. Penttilä (1995). Cloning and expression in *Saccharomyces cerevisiae* of a *Trichoderma reesei* beta-mannanase gene containing a cellulose binding domain. *Appl Environ Microbiol* **61**(3): 1090–1097.

Starr, M. P. and F. Moran (1962). Eliminative split of pectic substances by phyto-pathogenic soft-rot bacteria. *Science* **135**(3507): 920–921.

Staudinger, H. (1961). *From Organic Chemistry to Macromolecules*. New York: Wiley-Interscience.

Steenbakkers, P. J. M., H. R. Harhangi, M. W. Bosscher, M. M. C. van der Hooft, J. T. Keltjens, C. van der Drift, G. D. Vogels and H. J. M. op den Camp (2003). Beta-Glucosidase in cellulosome of the anaerobic fungus *Piromyces* sp. strain E2 is a family 3 glycoside hydrolase. *Biochem J* **370**(3): 963–970.

Steenbakkers, P. J. M., X.-L. Li, E. A. Ximenes, J. G. Arts, H. Chen, L. G. Ljungdahl and H. J. M. Op den Camp (2001). Noncatalytic docking domains of cellulo-somes of anaerobic fungi. *J Bacteriol* **183**(18): 5325–5333.

Steffler, F., J.-K. Guterl and V. Sieber (2013). Improvement of thermostable aldehyde dehydrogenase by directed evolution for application in synthetic cascade biomanufacturing. *Enzyme Microb Technol* **53**(5): 307–314.

Stemmer, W. P. C. (1994a). Rapid evolution of a protein in vitro by DNA shuffling. *Nature* **370**(6488): 389–391.

Stemmer, W. P. C. (1994b). DNA shuffling by random fragmentation and reassem-bly: In vitro recombination for molecular evolution. *Proc Natl Acad Sci USA* **91**(22): 10747–10751.

Stenberg, K., C. Tengborg, M. Galbe and G. Zacchi (1998). Optimisation of steam pretreatment of SO₂-impregnated mixed softwoods for ethanol production. *J Chem Technol Biotechnol* **71**(4): 299–308.

Stephen, C. F. (1989). The structure and functions of xyloglucan. *J Exp Bot* **40**: 1–11.

Stewart, G. G. and I. Russell (1987). Biochemical and genetic control of sugar and carbohydrate metabolism in yeasts. In *Yeast Biotechnology*, D. R. Berry, I. Russell and G. G. Stewart (Eds.), pp. 277–310. London: Allen and Unwin.

Stout, A. B. and E. J. Schreiner (1933). Results of a project in hybridizing poplars. *J Hered* **24**(6): 217–229.

Stricker, A. R., R. L. Mach and L. H. de Graaff (2008). Regulation of transcription of cellulases- and hemicellulases-encoding genes in *Aspergillus niger* and *Hypocrea jecorina* (*Trichoderma reesei*). *Appl Microbiol Biotechnol* **78**(2): 211–220.

Strobel, H. J., F. C. Caldwell and K. A. Dawson (1995). Carbohydrate transport by the anaerobic thermophile *Clostridium thermocellum* LQRI. *Appl Environ Microbiol* **61**(11): 4012–4015.

Stubbendieck, J., S. L. Hatch and C. Butterfield (1991). *North American Range Plants*. Lincoln, NE: University of Nebraska Press.

Stutzenberger, F. (1986). Hydrolysis products inhibit adsorption of *Trichoderma reesei* C30 cellulases to protein-extracted lucerne fibres. *Enzyme Microb Technol* **8**(6): 341–344.

Subhedar, P. B., N. R. Babu and P. R. Gogate (2015). Intensification of enzymatic hydrolysis of waste newspaper using ultrasound for fermentable sugar production. *Ultrason Sonochem* **22**: 326–332.

Sugano, Y. (2009). DyP-type peroxidases comprise a novel heme peroxidase family. *Cell Mol Life Sci* **66**(8): 1387–1403.

Sugano, Y., R. Muramatsu, A. Ichiyanagi, T. Sato and M. Shoda (2007). DyP, a unique dye-decolorizing peroxidase, represents a novel heme peroxidase family: ASP171 replaces the distal histidine of classical peroxidases. *J Biol Chem* **282**(50): 36652–36658.

Sugano, Y., R. Nakano, K. Sasaki and M. Shoda (2000). Efficient heterologous expression in *Aspergillus* oryzae of a unique dye-decolorizing peroxidase, DyP, of Geotrichum candidum Dec 1. *Appl Environ Microbiol* **66**: 1754–1758.

Sugano, Y., K. Sasaki and M. Shoda (1999). cDNA cloning and genetic analysis of a novel decolorizing enzyme, peroxidase gene dyp from Geotrichum candidum Dec 1. *J Biosci Bioeng* **87**(4): 411–417.

Sun, N., M. Rahman, Y. Qin, M. L. Maxim, H. Rodriguez and R. D. Rogers (2009). Complete dissolution and partial delignification of wood in the ionic liquid 1-ethyl-3-methylimidazolium acetate. *Green Chem* **11**(5): 646–655.

Sun, R., X. F. Sun, S. Q. Wang, W. Zhu and X. Y. Wang (2002). Ester and ether linkages between hydroxycinnamic acids and lignins from wheat, rice, rye, and barley straws, maize stems, and fast-growing poplar wood. *Ind Crop Prod* **15**: 179–188.

Sun, X. F., R. Sun, P. Fowler and M. S. Baird (2005). Extraction and characterization of original lignin and hemicelluloses from wheat straw. *J Agric Food Chem* **53**(4): 860–870.

Sunna, A. and G. Antranikian (1997). Xylanolytic enzymes from fungi and bacteria. *Crit Rev Biotechnol* **17**(1): 39–67.

Suye, S. (1997). Purification and properties of alcohol oxidase from Candida methanosorbosa M-2003. *Curr Microbiol* **34**(6): 374–377.

Sygmund, C., D. Kracher, S. Scheiblbrandner, K. Zahma, A. K. Felice, W. Harreither, R. Kittl and R. Ludwig (2012). Characterization of the two *Neurospora crassa* cellobiose dehydrogenases and their connection to oxidative cellulose degradation. *Appl Environ Microbiol* **78**(17): 6161–6171.

Takano, M., M. Nakamura and M. Yamaguchi (2010). Glyoxal oxidase supplies hydrogen peroxide at hyphal tips and on hyphal wall to manganese peroxidase of white-rot fungus *Phanerochaete crassa* WD1694. *J Wood Sci* **56**(4): 307–313.

Takashima, S., A. Nakamura, M. Hidaka, H. Masaki and T. Uozumi (1999). Molecular cloning and expression of the novel fungal β-glucosidase genes from *Humicola grisea* and *Trichoderma reesei*. *J Biochem* **125**(4): 728–736.

Talmadge, K. W., K. Keegstra, W. D. Bauer and P. Albersheim (1973). The structure of plant cell walls I. The macromolecular components of the walls of suspension-cultured sycamore cells with a detailed analysis of the pectic polysaccharides. *Plant Physiol* **51**(1): 158–173.

Tamaru, Y. and R. H. Doi (2001). Pectate lyase A, an enzymatic subunit of the *Clostridium cellulovorans* cellulosome. *Proc Natl Acad Sci USA* **98**(7): 4125–4129.

Tan, S. S. Y., D. R. MacFarlane, J. Upfal, L. A. Edye, W. O. S. Doherty, A. F. Patti, J. M. Pringle and J. L. Scott (2009). Extraction of lignin from lignocellulose at atmospheric pressure using alkylbenzenesulfonate ionic liquid. *Green Chem* **11**(3): 339–345.

Taniai, T., Sukuragawa, A., Okutani, T. (2001). Fluorometric determination of ethanol in liquor samples by flow-injection analysis using an immobilized enzyme-reactor column with packing prepared by coupling alcohol oxidase and peroxidase onto chitosan beads. *J AOAC Int* **84**(5): 1475–1483.

Tao, H. and V. W. Cornish (2002). Milestones in directed enzyme evolution. *Curr Opin Chem Biol* **6**(6): 858–864.

Tassinari, T., C. Macy and L. Spano (1980). Energy-requirements and process design considerations in compression-milling pretreatment of cellulosic wastes for enzymatic-hydrolysis. *Biotechnol Bioeng Symp* **22**(8): 1689–1705.

Tassinari, T. H., C. F. Macy and L. A. Spano (1982). Technology advances for continuous compression milling pretreatment of lignocellulosics for enzymatic hydrolysis. *Biotechnol Bioeng* **24**(7): 1495–1505.

Taylor, S. V., P. Kast and D. Hilvert (2001). Investigating and engineering enzymes by genetic selection. *Angew Chem Int Ed* **40**(18): 3310–3335.

Teeri, T., I. Salovuori and J. Knowles (1983). The molecular cloning of the major cellulase gene from *Trichoderma Reesei*. *Nat Biotech* **1**(8): 696–699.

Teeri, T. T., P. Lehtovaara, S. Kauppinen, I. Salovuori and J. Knowles (1987). Homologous domains in *Trichoderma reesei* cellulolytic enzymes: Gene sequence and expression of cellobiohydrolase II. *Gene* **51**(1): 43–52.

Teleman, A., M. Tenkanen, A. Jacobs and O. Dahlman (2002). Characterization of O-acetyl-(4-O-methylglucurono)xylan isolated from birch and beech. *Carbohydr Res* **337**(4): 373–377.

Temp, U. and C. Eggert (1999). Novel interaction between laccase and cellobiose dehydrogenase during pigment synthesis in the white rot fungus *Pycnoporus cinnabarinus*. *Appl Environ Microbiol Mol Biol Rev* **65**(2): 389–395.

Tenkanen, M., J. Puls and K. Poutanen (1992). Two major xylanases of *Trichoderma reesei*. *Enzyme Microb Technol* **14**(7): 566–574.

Tenkanen, M. and M. Siika-aho (2000). An alpha-glucuronidase of *Schizophyllum commune* acting on polymeric xylan. *J Biotechnol* **78**(2): 149–161.

Tenkanen, M., M. Vršanská, M. Siika-aho, D. W. Wong, V. Puchart, M. Penttilä, M. Saloheimo and P. Biely (2013). Xylanase XYN IV from *Trichoderma reesei* showing exo- and endo-xylanase activity. *FEBS J* **280**(1): 285–301.

Teugjas, H. and P. Valjamae (2013). Product inhibition of cellulases studied with 14C-labeled cellulose substrates. *Biotechnol Biofuels* **6**(1): 104.

Teymouri, F., L. Laureano-Perez, H. Alizadeh and B. E. Dale (2005). Optimization of the ammonia fiber explosion (AFEX) treatment parameters for enzymatic hydrolysis of corn stover. *Bioresour Technol* **96**(18): 2014–2018.

Teze, D., F. Daligault, V. Ferrières, Y. H. Sanejouand and C. Tellier (2015). Semi-rational approach for converting a GH36 α-glycosidase into an α-transglycosidase. *Glycobiology* **25**(4): 420–427.

Tharanathan, R. N., G. C. Reddy, G. G. Muralikrishna, N. S. Susheelamma and U. Ramadas Bhat (1994). Structure of a galactoarabinan-rich pectic polysaccharide of native and fermented blackgram (*Phaseolus mungo*). *Carbohydr Polym* **23**(2): 121–127.

The Regents of the University of California (2015). *Trichoderma reesei* genome database v2.0. Retrieved July, 16, 2015, from http://genome.jgi-psf.org/Trire2/Trire2.home.html.

Themelis, N. J. (2003). An overview of the global waste-to-energy industry. *Waste Manage World* **3**: 40–47.

Thompson, J. E. and S. C. Fry (2000). Evidence for covalent linkage between xyloglucan and acidic pectins in suspension-cultured rose cells. *Planta* **211**(2): 275–286.

Tien, M. and T. K. Kirk (1983). Lignin-degrading enzyme from the Hymenomycete *Phanerochaete chrysosporium* burds. *Science* **221**(4611): 661–663.

Tien, M. and T. K. Kirk (1984). Lignin-degrading enzyme from *Phanerochaete chrysosporium*: Purification, characterization, and catalytic properties of a unique H2O2-requiring oxygenase. *Proc Natl Acad Sci USA* **81**(8): 2280–2284.

Tilbeurgh, H. V., P. Tomme, M. Claeyssens, R. Bhikhabhai and G. Pettersson (1986). Limited proteolysis of the cellobiohydrolase I from *Trichoderma reesei*: Separation of functional domains. *Febs Lett* **204**(2): 223–227.

Timell, T. E. (1960). Isolation and Properties of an O-Acetyl-4-O-methylglucuronoxyloglycan from the Wood of White Birch (*Betula papyrifera*). *J Am Chem Soc* **82**(19): 5211–5215.

Timell, T. E. (1967). Recent progress in the chemistry of wood hemicelluloses. *Wood Sci Technol* **1**(1): 45–70.

Tiné, M. A. S., C. O. Silva, D. U. de Lima, N. C. Carpita and M. S. Buckeridge (2006). Fine structure of a mixed-oligomer storage xyloglucan from seeds of *Hymenaea courbaril*. *Carbohydr Polym* **66**(4): 444–454.

Tishkov, V. I., A. V. Gusakov, A. S. Cherkashina and A. P. Sinitsyn (2013). Engineering the pH-optimum of activity of the GH12 family endoglucanase by site-directed mutagenesis. *Biochimie* **95**(9): 1704–1710.

Tiwari, M. K., R. Singh, R. K. Singh, I. W. Kim and J. K. Lee (2012). Computational approaches for rational design of proteins with novel functionalities. *Comput Struct Biotechnol J* **2**(3): 1–13.

Tokatlidis, K., S. Salamitou, P. Béguin, P. Dhurjati and J.-P. Aubert (1991). Interaction of the duplicated segment carried by *Clostridium thermocellum* cellulases with cellulosome components. *FEBS Lett* **291**(2): 185–188.

Tokuriki, N. and D. S. Tawfik (2009). Stability effects of mutations and protein evolvability. *Curr Opin Struct Biol* **19**(5): 596–604.

Tomme, P., A. Boraston, B. McLean, J. Kormos, L. Creagh, K. Sturch, N. R. Gilkes, C. A. Haynes, R. A. J. Warren and D. G. Kilburn (1998a). Characterization and affinity applications of cellulose-binding domains. *J Chromatogr B* **715**(1): 283–296.

Tomme, P., H. Van Tilbeurgh, G. Pettersson, J. Van Damme, J. Vandekerckhove, J. Knowles, T. Teeri and M. Claeyssens (1988b). Studies of the cellulolytic system of *Trichoderma reesei* QM 9414. Analysis of domain function in two cellobiohydrolases by limited proteolysis. *Eur J Biochem* **170**(3): 575–581.

Tormo, J., R. Lamed, A. J. Chirino, E. Morag, E. A. Bayer, Y. Shoham and T. A. Steitz (1996). Crystal structure of a bacterial family-III cellulose-binding domain: A general mechanism for attachment to cellulose. *EMBO J* **15**(21): 5739–5751.

Torronen, A., R. L. Mach, R. Messner, R. Gonzalez, N. Kalkkinen, A. Harkki and C. P. Kubicek (1992). The two major xylanases from *Trichoderma Reesei*: Characterization of both enzymes and genes. *Nat Biotech* **10**(11): 1461–1465.

Travaini, R., M. D. M. Otero, M. Coca, R. Da-Silva and S. Bolado (2013). Sugarcane bagasse ozonolysis pretreatment: Effect on enzymatic digestibility and inhibitory compound formation. *Bioresour Technol* **133**: 332–339.

Trojanowski, J., A. Leonowicz and B. Hampel (1966). Exoenzymes in fungi degrading lignin. II. Demethoxylation of lignin and vanillic acid. *Acta Microbiol Pol* **15**(1): 17–22.

Trudeau, D. L., T. M. Lee and F. H. Arnold (2014). Engineered thermostable fungal cellulases exhibit efficient synergistic cellulose hydrolysis at elevated temperatures. *Biotechnol Bioeng* **111**(12): 2390–2397.

Tuka, K., V. V. Zverlov and G. A. Velikodvorskaya (1992). Synergism between *Clostridium thermocellum* cellulases cloned in *Escherichia coli*. *Appl Biochem Biotechnol* **37**(2): 201–207.

Tuka, K., V. V. Zverlov, B. K. Bumazkin, G. A. Velikodvorskaya and A. Y. Strongin (1990). Cloning and expression of *Clostridium thermocellum* genes coding for thermostable exoglucanases (cellobiohydrolases) in *Escherichia coli* cells. *Biochem Biophys Res Commun* **169**(3): 1055–1060.

Tuor, U., H. Wariishi, H. E. Schoemaker and M. H. Gold (1992). Oxidation of phenolic arylglycerol β-aryl ether lignin model compounds by manganese peroxidase from *Phanerochaete chrysosporium*: Oxidative cleavage of an α-carbonyl model compound. *Biochemistry* **31**(21): 4986–4995.

Turner, M. B., S. K. Spear, J. G. Huddleston, J. D. Holbrey and R. D. Rogers (2003). Ionic liquid salt-induced inactivation and unfolding of cellulase from *Trichoderma reesei*. *Green Chem* **5**(4): 443–447.

Turner, N. J. (2009). Directed evolution drives the next generation of biocatalysts. *Nat Chem Biol* **5**(8): 567–573.

Tyson, K. S., M. Rymes and E. Hammond (1996). Future potential for MSW energy development. *Biomass Bioenergy* **10**(2–3): 111–124.

Umezawa, T., S. Kawai, S. Yokota and T. Higuchi (1986). Aromatic ring cleavage of various beta-0-4 lignin model dimers by *Phanerochaete chrysosporium*. *Wood Res* **73**: 8–17.

United States Environment Protection Agency (2007). *Greenhouse Gas Impacts of Expanded Renewable and Alternative Fuels Use*. Washington, DC: Office of Transportation and Air Quality.

United States Environmental Protection Agency (2014). Municipal solid waste generation, recycling, and disposal in the United States: Facts and figures for 2012.

USDA Foreign Agricultural Service (2014). *Sugar: World Markets and Trade*.

Ustinov, B. B., A. V. Gusakov, A. I. Antonov and A. P. Sinitsyn (2008). Comparison of properties and mode of action of six secreted xylanases from *Chrysosporium lucknowense*. *Enzyme Microb Technol* **43**(1): 56–65.

Usui, K., K. Ibata, T. Suzuki and K. Kawai (1999). XynX, a possible exo-xylanase of Aeromonas caviae ME-1 that produces exclusively xylobiose and xylotetraose from xylan. *Biosci Biotechnol Biochem* **63**(8): 1346–1352.

Utt, E. A., C. K. Eddy, K. F. Keshav and L. O. Ingram (1991). Sequencing and expression of the Butyrivibrio fibrisolvens xylB gene encoding a novel bifunctional protein with beta-D-xylosidase and alpha-L-arabinofuranosidase activities. *Appl Environ Microbiol* **57**(4): 1227–1234.

Uzan, E., P. Nousiainen, V. Balland, J. Sipila, F. Piumi, D. Navarro, M. Asther, E. Record and A. Lomascolo (2010). High redox potential laccases from the ligninolytic fungi *Pycnoporus coccineus* and *Pycnoporus sanguineus* suitable for white biotechnology: From gene cloning to enzyme characterization and applications. *J Appl Microbiol* **108**(6): 2199–2213.

Uziie, M., M. Matsuo and T. Yasui (1985). Possible identity of β-Xylosidase and β-Glucosidase of *Chaetomium trilaterale*. *Agric Biol Chem* **49**(4): 1167–1173.

Vaaje-Kolstad, G., B. Westereng, S. J. Horn, Z. Liu, H. Zhai, M. Sorlie and V. G. H. Eijsink (2010). An oxidative enzyme boosting the enzymatic conversion of recalcitrant polysaccharides. *Science* **330**(6001): 219–222.

Vadas, P. A., K. H. Barnett and D. J. Undersander (2008). Economics and energy of ethanol production from alfalfa, corn, and switchgrass in the upper midwest, USA. *BioEnergy Res* **1**(1): 44–55.

Valent, B. S. and P. Albersheim (1974). The structure of plant cell walls: v. On the binding of xyloglucan to cellulose fibers. *Plant Physiol* **54**(1): 105–108.

Valenzuela, S. V., C. Valls, M. B. Roncero, T. Vidal, P. Diaz and F. I. J. Pastor (2013). Effectiveness of novel xylanases belonging to different GH families on lignin and hexenuronic acids removal from specialty sisal fibres. *J Chem Technol Biotechnol* **89**(3): 401–406.

van den Brink, J. and R. P. de Vries (2011). Fungal enzyme sets for plant polysaccharide degradation. *Appl Microbiol Biotechnol* **91**(6): 1477–1492.

van der Vlugt-Bergmans, C. J., P. J. Meeuwsen, A. G. Voragen and A. J. van Ooyen (2000). Endo-xylogalacturonan hydrolase, a novel pectinolytic enzyme. *Appl Environ Microbiol* **66**(1): 36–41.

Van Doorslaer, E., H. Kersters-Hilderson and C. K. De Bruyne (1985). Hydrolysis of β-d-xylo-oligosaccharides by β-d-xylosidase from *Bacillus pumilus*. *Carbohydr Res* **140**(2): 342–346.

Van Heiningen, A. (2006). Converting a kraft pulp mill into an integrated forest biorefinery. *Pulp Paper Can* **107**(6): 38–43.

van Walsum, G. P. and H. Shi (2004). Carbonic acid enhancement of hydrolysis in aqueous pretreatment of corn stover. *Bioresour Technol* **93**(3): 217–226.

Varner, J. E. and L.-S. Lin (1989). Plant cell wall architecture *Cell* **56**: 231–239.

Vasiltsova, T. V., S. P. Verevkin, E. Bich, A. Heintz, R. Bogel-Lukasik and U. Domanska (2004). Thermodynamic properties of mixtures containing ionic liquids. Activity coefficients of ethers and alcohols in 1-methyl-3-ethylimidazolium bis(trifluoromethyl-sulfonyl) imide using the transpiration method. *J Chem Eng Data* **50**(1): 142–148.

Vassilev, S. V., D. Baxter, L. K. Andersen, C. G. Vassileva and T. J. Morgan (2012). An overview of the organic and inorganic phase composition of biomass. *Fuel* **94**: 1–33.

Veitch, N. C. (2004). Horseradish peroxidase: A modern view of a classical enzyme. *Phytochemistry* **65**(3): 249–259

Verbruggen, M. A., G. Beldman and A. G. Voragen (1998a). Enzymic degradation of sorghum glucuronoarabinoxylans leading to tentative structures. *Carbohydr Res* **306**(1–2): 275–282.

Verbruggen, M. A., B. A. Spronk, H. A. Schols, G. Beldman, A. G. J. Voragen, J. R. Thomas, J. P. Kamerling and J. F. G. Vliegenthart (1998b). Structures of enzymically derived oligosaccharides from sorghum glucuronoarabinoxylan. *Carbohydr Res* **306**(1–2): 265–274.

Verma, O. P., A. Singh, N. Singh and O. Chaudhary (2011a). Isolation, purification and characterization of β-glucosidase from Rauvolfia serpentina. *J Chem Eng Process Technol* **2**(5): 119.

Verma, P., T. Watanabe, Y. Honda and T. Watanabe (2011b). Microwave-assisted pretreatment of woody biomass with ammonium molybdate activated by H2O2. *Bioresour Technol* **102**(4): 3941–3945.

Vierhuis, E., H. A. Schols, G. Beldman and A. G. J. Voragen (2000). Isolation and characterisation of cell wall material from olive fruit (Olea europaea cv koroneiki) at different ripening stages. *Carbohydr Polym* **43**(1): 11–21.

Viña-Gonzalez, J., D. Gonzalez-Perez, P. Ferreira, A. T. Martinez and M. Alcalde (2015). Focused directed evolution of aryl-alcohol oxidase in yeast using chimeric signal peptides. *Appl Environ Microbiol* **81**: 6451–6462.

Vincken, J. P., H. A. Schols, R. J. Oomen, M. C. McCann, P. Ulvskov, A. G. Voragen and R. G. Visser (2003). If homogalacturonan were a side chain of rhamnogalacturonan I. Implications for cell wall architecture. *Plant Physiol* **132**(4): 1781–1789.

Visser, F., B. Muller, J. Rose, D. Prufer and G. A. Noll (2016). Forizymes-functionalized artificial forisomes as platforms for the production and immobilization of single enzymes and multi-enzyme complexes. *Nat Sci Rep* **6**: 30839.

Vitali, J., B. Schick, H. C. M. Kester, J. Visser and F. Jurnak (1998). The three-dimensional structure of *Aspergillus niger* pectin lyase B at 1.7-Å resolution. *Plant Physiol* **116**(1): 69–80.

Vlasenko, E., M. Schülein, J. Cherry and F. Xu (2010). Substrate specificity of family 5, 6, 7, 9, 12, and 45 endoglucanases. *Bioresour Technol* **101**(7): 2405–2411.

Vocadlo, D. J. and G. J. Davies (2008). Mechanistic insights into glycosidase chemistry. *Curr Opin Chem Biol* **12**: 539–555.

Voelker, S. L., B. Lachenbruch, F. C. Meinzer and S. H. Strauss (2011). Reduced wood stiffness and strength, and altered stem form, in young antisense 4CL transgenic poplars with reduced lignin contents. *New Phytol* **189**(4): 1096–1109.

Volc, J., P. Sedmera and V. Musílek (1978). Glucose-2-oxidase activity and accumulation of D-*arabino*-2-hexosulose in cultures of the basidiomycete *Oudemansiella mucida*. *Folia Microbiol (Praha)* **23**(4): 292–298.

von Gal Milanezi, N., D. P. Gómez Mendoza, F. Gonçalves de Siqueira, L. Paulino Silva, C. A. Ornelas Ricart and E. X. Ferreira Filho (2012). Isolation and characterization of a xylan-degrading enzyme from *Aspergillus niger* van Tieghem LPM 93 with potential for industrial applications. *Bioenerg Res* **5**(2): 363–371.

Vu, V. H. and K. Kim (2012). Improvement of cellulase activity using error-prone rolling circle amplification and site-directed mutagenesis. *J Microbiol Biotechnol* **22**(5): 607–613.

Vu, V. V., W. T. Beeson, E. A. Span, E. R. Farquhar and M. A. Marletta (2014). A family of starch-active polysaccharide monooxygenases. *Proc Natl Acad Sci USA* **111**(38): 13822–13827.

Wahler, D. and J.-L. Reymond (2001). Novel methods for biocatalyst screening. *Curr Opin Chem Biol* **5**(2): 152–158.

Wanderleya, M. C. A., C. Martín, G. J. M. Rocha and E. R. Gouveia (2013). Increase in ethanol production from sugarcane bagasse based on combined pretreatments and fed-batch enzymatic hydrolysis. *Bioresour Technol* **128**: 448–453.

Wang, C., K. Zhang, C. Zhongjun, H. Cai, W. Honggui and P. Ouyang (2015). Directed evolution and mutagenesis of lysine decarboxylase from *Hafnia alvei* AS1.1009 to improve its activity toward efficient cadaverine production. *Biotechnol Bioprocess Eng* **20**(3): 439–446.

Wang, H., H. Wang, S. Shi, J. Duan and S. Wang (2012a). Structural characterization of a homogalacturonan from *Capparis spinosa* L. fruits and anti-complement activity of its sulfated derivative. *Glycoconjugate J* **29**(5–6): 379–387.

Wang, H., J. Wang, Z. Fang, X. Wang and H. Bu (2010). Enhanced bio-hydrogen production by anaerobic fermentation of apple pomace with enzyme hydrolysis. *Int J Hydrogen Energy* **35**(15): 8303–8309.

Wang, H., G. Wei, F. Liu, G. Banerjee, M. Joshi, S. W. A. Bligh, S. Shi et al. (2014a). Characterization of two homogalacturonan pectins with immunomodulatory activity from green tea. *Int J Mol Sci* **15**(6): 9963–9978.

Wang, H.-C., Y.-C. Chen and R.-S. Hseu (2014b). Purification and characterization of a cellulolytic multienzyme complex produced by *Neocallimastix patriciarum* J11. *Biochem Biophys Res Commun* **451**(2): 190–195.

Wang, J., Q. Zhang, Z. Huang and Z. Liu (2013). Directed evolution of a family 26 glycoside hydrolase: Endo-beta-1,4-mannanase from *Pantoea agglomerans* A021. *J Biotechnol* **167**(3): 350–356.

Wang, L., Y. Zhang and P. Gao (2008). A novel function for the cellulose binding module of cellobiohydrolase I. *Sci China C Life Sci* **51**(7): 620–629.

Wang, Q. Q., Z. He, Z. Zhu, Y. H. Zhang, Y. Ni, X. L. Luo and J. Y. Zhu (2012b). Evaluations of cellulose accessibilities of lignocelluloses by solute exclusion and protein adsorption techniques. *Biotechnol Bioeng* **109**(2): 381–389.

Wang, T., X. Liu, Q. Yu, X. Zhang, Y. Qu, P. Gao and T. Wang (2005). Directed evolution for engineering pH profile of endoglucanase III from *Trichoderma reesei*. *Biomol Eng* **22**(1–3): 89–94.

Wang, T., O. Zabotina and M. Hong (2012c). Pectin–cellulose interactions in the *Arabidopsis* primary cell wall from two-dimensional magic-angle-spinning solid-state nuclear magnetic resonance. *Biochemistry* **51**: 9846–9856.

Wang, W., T. Yuan, K. Wang, B. Cui and Y. Dai (2012d). Combination of biological pretreatment with liquid hot water pretreatment to enhance enzymatic hydrolysis of Populus tomentosa. *Bioresour Technol* **107**: 282–286.

Wang, W. K., K. Kruus, J. H. Wu (1993). Cloning and DNA sequence of the gene coding for *Clostridium thermocellum* cellulase Ss (CelS), a major cellulosome component. *J Bacteriol* **175**(5): 1293–1302.

Wang, W. K. and J. H. D. Wu (1993). Structural features of the *Clostridium thermocellum* cellulase Ss gene. *Appl Biochem Biotechnol* **39– 40**(1): 149–158.

Wang, X. J., Y. J. Peng, L. Q. Zhang, A. N. Li and D. C. Li (2012e). Directed evolution and structural prediction of cellobiohydrolase II from the thermophilic fungus *Chaetomium thermophilum*. *Appl Microbiol Biotechnol* **95**(6): 1469–1478.

Wang, Y., R. Tang, J. Tao, Z. Wang, B. Zheng and Y. Feng (2012f). Chimeric cellulase matrix for investigating intramolecular synergism between non-hydrolytic disruptive functions of carbohydrate-binding modules and catalytic hydrolysis. *J Biol Chem* **287**(35): 29568–29578.

Ward, G., Y. Hadar, I. Bilkis, C. G. Dosoretz (2003). Mechanistic features of lignin peroxidase-catalyzed oxidation of substituted phenols and 1,2-dimethoxyarenes. *J Biol Chem* **278**(41): 39726–39734.

Ward, G., Y. Hadar and C. G. Dosoretz (2001). Inactivation of lignin peroxidase during oxidation of the highly reactive substrate ferulic acid. *Enzyme Microb Technol* **29**(1): 34–41.

Wasserscheid, P. and W. Keim (2000). Ionic liquids-new solutions for transition metal catalysis. *Angew Chem* **39**(21): 3772–3789.

Watanabe, A., K. Hiraga, M. Suda, H. Yukawa and M. Inui (2015). Functional characterization of *Corynebacterium alkanolyticum* β-xylosidase and xyloside ABC transporter in *Corynebacterium glutamicum*. *Appl Environ Microbiol* **81**: 4173–4183.

Watanabe, T. and T. Koshijima (1988). Evidence for an ester linkage between lignin and glucuronic acid in lignin-carbohydrate complexes by DDQ-oxidation. *Agric Biol Chem* **52**(11): 2953–2955.

Watanabe, T., J. Ohnishi, Y. Yamasaki, S. Kaizu and T. Koshijima (1989). Binding sites of the ether linkages between lignin and hemicelluloses in lignin-carbohydrate complexes by DDQoxidation. *Agric Biol Chem* **53**: 2233–2252.

Watkins, K. W. (1983). Heating in microwave ovens: An example of dipole moments in action. *J Chem Educ* **60**(12): 1043.

Watson, B. J., H. Zhang, A. G. Longmire, Y. H. Moon and S. W. Hutcheson (2009). Processive endoglucanases mediate degradation of cellulose by Saccharophagus degradans. *J Bacteriol* **191**(18): 5697–5705.

Weaver, J. E. (1968). *Prairie Plants and their Environment: A Fifty-year Study in the Midwest*. Lincoln, NE, University of Nebraska Press, p. 276.

Wei, Y. D., S. J. Lee, K. S. Lee, Z. Z. Gui, H. J. Yoon, I. Kim, Y. H. Je, X. Guo, H. D. Sohn and B. R. Jin (2005). N-glycosylation is necessary for enzymatic activity of a beetle (*Apriona germari*) cellulase. *Biochem Biophys Res Commun* **329**(1): 331–336.

Weil, J., M. Brewer, R. Hendrickson, A. Sarikaya and M. R. Ladisch (1998). Continuous pH monitoring during pretreatment of yellow poplar wood sawdust by pressure cooking in water. *Appl Biochem Biotechnol* **70– 72**(1): 99–111.

Wells, J. A. (1990). Additivity of mutational effects in proteins. *Biochemistry* **29**(37): 8509–8517.

Wende, G. and S. C. Fry (1997). 2-O-β-d-xylopyranosyl-(5-O-feruloyl)-l-arabinose, a widespread component of grass cell walls. *Phytochemistry* **44**(6): 1019–1030.

Westermark, U. and K.-E. Eriksson (1974a). Carbohydrate-dependent enzymatic quinone reduction during lignin degradation. *Acta Chem Scand* **28b**(2): 204–208.

Westermark, U. and K.-E. Eriksson (1974b). Cellobiose-quinone oxidoreductase, a new wood-degrading enzyme from white rot fungi. *Acta Chem Scand* **28b**: 209–214.

Wilkie, A. C., K. J. Riedesel and J. M. Owens (2000). Stillage characterization and anaerobic treatment of ethanol stillage from conventional and cellulosic feedstocks. *Biomass Bioenergy* **19**(2): 63–102.

Wilkie, K. C. B. (1979). The hemicelluloses of grasses and cereals. In *Advances in Carbohydrate Chemistry and Biochemistry*, R. S. Tipson and H. Derek (Eds.), Vol. 36, pp. 215–264. Salt Lake City, UT: Academic Press.

Willats, W. G., L. McCartney, W. Mackie and J. P. Knox (2001). Pectin: Cell biology and prospects for functional analysis. *Plant Mol Biol* **47**(1–2): 9–27.

Willfor, S., R. Sjöholm, C. Laine, M. Roslund, J. Hemming and B. Holmbom (2003). Characterisation of water-soluble galactoglucomannans from Norway spruce wood and thermomechanical pulp. *Carbohydr Polym* **52**: 175–187.

Williams, J. S., R. Hoos and S. G. Withers (2000). Nanomolar versus millimolar inhibition by xylobiose-derived azasugars: Significant differences between two structurally distinct xylanases. *J Am Chem Soc* **122**: 2223–2234.

Williams, J. T., J. DransVeld, P. M. Ganapathy, W. Liese, S. M. Nor and C. B. Sastry (1991). Research needs for bamboo and rattan to the year 2000. Singapore: International Fund for Agricultural Reasearch/International Network for Bamboo and Rattan.

Williamson, G. (1991). Purification and characterization of pectin acetylesterase from orange peel. *Phytochemistry* **30**: 445–449.

Williamson, G., C. B. Faulds, J. A. Matthew, D. B. Archer, V. J. Morris, G. J. Brownsey and M. J. Ridout (1990). Gelation of sugarbeet and citrus pectins using enzymes extracted from orange peel. *Carbohydr Polym* **13**(4): 387–397.

Wilson, C. A. and T. M. Wood (1992). The anaerobic fungus *Neocallimastix frontalis*: Isolation and properties of a cellulosome-type enzyme fraction with the capacity to solubilize hydrogen-bond-ordered cellulose *Appl Microbiol Biotechnol* **37**: 125–129.

Wilson, D. B. (2008). Aerobic microbial cellulase systems. In *Biomass Recalcitrance: Deconstructing the Plant Cell Wall for Bioenergy*, M. E. Himmel (Ed.), pp. 374–392. Oxford, UK: Blackwell Publishing.

Wising, U. and P. Stuart (2006). Identifying the Canadian forest biorefinery. *Pulp Paper Can* **107**(6): 25–30.

Wong, D. W. (2009). Structure and action mechanism of ligninolytic enzymes. *Appl Biochem Biotechnol* **157**(2): 174–209.

Wood, J. D. and P. M. Wood (1992). Evidence that cellobiose:quinone oxidoreductase from *Phanerochaete chrysosporium* is a breakdown product of cellobiose oxidase. *Biochim Biophys Acta* **1119**(1): 90–96.

Wood, T. M. and S. I. McCrae (1972). The purification and properties of the C1 component of *Trichoderma koningii* cellulase. *Biochem J* **128**(5): 1183–1892.

Wood, T. M. and S. I. McCrae (1978). The cellulase of *Trichoderma koningii*. Purification and properties of some endoglucanase components with special reference to their action on cellulose when acting alone and in synergism with the cellobiohydrolase. *Biochem J* **171**(1): 61–72.

Wood, T. M. and S. I. McCrae (1979). Synergism between enzymes involved in the solubilization of native cellulose. *Adv Chem Ser* **181**: 181–209.

Wooley, R., M. Ruth, D. Glassner and J. Sheehan (1999). Process design and costing of bioethanol technology: A tool for determining the status and direction of research and development. *Biotechnol Prog* **15**(5): 794–803.

Worth, C. L., R. Preissner and T. L. Blundell (2011). SDM—A server for predicting effects of mutations on protein stability and malfunction. *Nucleic Acids Res* **39**(2): W215–222.

Wu, I. and F. H. Arnold (2013). Engineered thermostable fungal Cel6A and Cel7A cellobiohydrolases hydrolyze cellulose efficiently at elevated temperatures. *Biotechnol Bioeng* **110**(7): 1874–1883.

Wyman, C. E. (1994). Ethanol from lignocellulosic biomass: Technology, economics, and opportunities. *Bioresour Technol* **50**(1): 3–16.

Wyman, C. E. (1996). *Handbook on Bioethanol: Production and Utilization*. Washington, DC: Taylor & Francis Group.

Wyman, C. E., B. E. Dale, R. T. Elander, M. Holtzapple, M. R. Ladisch, Y. Y. Lee, C. Mitchinson and J. N. Saddler (2009). Comparative sugar recovery and fermentation data following pretreatment of poplar wood by leading technologies. *Biotechnol Prog* **25**(2): 333–339.

Xiao, Z., H. Bergeron, S. Grosse, M. Beauchemin, M.-L. Garron, D. Shaya, T. Sulea, M. Cygler and P. C. K. Lau (2008). Improvement of the thermostability and activity of a pectate lyase by single amino acid substitutions, using a strategy based on melting-temperature-guided sequence alignment. *Appl Environ Microbiol* **74**(4): 1183–1189.

Xiao, Z., X. Zhang, D. J. Gregg and J. N. Saddler (2004). Effects of sugar inhibition on cellulases and β-glucosidase during enzymatic hydrolysis of softwood substrates. *Appl Biochem Biotechnol* **113– 116**(1–3): 1115–1126.

Xiaochen, W., Li, M., Li, Z., Lv, L., Zhang, Y., Li, C. (2016). Amyloid-graphene oxide as immobilization platform of Au nanocatalysts and enzymes for improved glucose-sensing activity. *J Colloid Interface Sci* **490**: 336–342.

Xie, H., H. J. Gilbert, S. J. Charnock, G. J. Davies, M. P. Williamson, P. J. Simpson, S. Raghothama et al. (2001). *Clostridium thermocellum* Xyn10B carbohydrate-binding module 22-2: The role of conserved amino acids in ligand binding. *Biochemistry* **40**(31): 9167–9176.

Xiong, H., O. Turunen, O. Pastinen, M. Leisola and N. von Weymarn (2004). Improved xylanase production by *Trichoderma reesei* grown on L-arabinose and lactose or D-glucose mixtures. *Appl Microbiol Biotechnol* **64**(3): 353–358.

Xiong, J., J. Ye, W. Z. Liang and P. M. Fan (2000). Influence of microwave on the ultrastructure of cellulose I. *J South China Univ Technol* **28**: 84–89.

Xiong, J.-S., M. Balland-Vanney, Z.-P. Xie, M. Schultze, A. Kondorosi, E. Kondorosi and C. Staehelin (2007). Molecular cloning of a bifunctional β-xylosidase/α-L-arabinosidase from alfalfa roots: Heterologous expression in *Medicago truncatula* and substrate specificity of the purified enzyme. *J Exp Bot* **58**(11): 2799–2810.

Xu, F. and H. Ding (2007). A new kinetic model for heterogeneous (or spatially confined) enzymatic catalysis: Contributions from fractal and jamming (overcrowding) effects. *Appl Catal A Gen* **317**: 70–81.

Xu, H., F. Zhang, H. Shang, X. Li, J. Wang, D. Qiao and Y. Cao (2013a). Alkalophilic adaptation of XynB endoxylanase from *Aspergillus niger* via rational design of p*Ka* of catalytic residues. *J Biosci Bioeng* **115**(6): 618–622.

Xu, J. and J. C. Smith (2010). Probing the mechanism of cellulosome attachment to the *Clostridium thermocellum* cell surface: Computer simulation of the Type II cohesin-dockerin complex and its variants. *Protein Eng Des Sel* **23**(10): 759–768.

Xu, J., M. H. Thomsen and A. B. Thomsen (2010). Ethanol production from hydrothermal pretreated corn stover with a loop reactor. *Biomass Bioenergy* **34**(3): 334–339.

Xu, J., N. Takakuwa, M. Nogawa, H. Okada and Y. Morikawa (1998). A third xylanase from *Trichoderma reesei* PC-3-7. *Appl Microbiol Biotechnol* **49**(6): 718–724.

Xu, M., R. Zhang, X. Liu, J. Shi, Z. Xu and Z. Rao (2013b). Improving the acidic stability of a β-mannanase from *Bacillus subtilis* by site-directed mutagenesis. *Process Biochem* **48**(8): 1166–1173.

Yagüe, E., P. Béguin and J. P. Aubert (1990). Nucleotide sequence and deletion analysis of the cellulase-eneoding gene celH of *Clostridium thermocellum*. *Gene* **89**(1): 61–67.

Yamada, R., T. Higo, C. Yoshikawa, H. China, M. Yasuda and H. Ogino (2015). Random mutagenesis and selection of organic solvent-stable haloperoxidase from *Streptomyces aureofaciens*. *Biotechnol Prog* **31**: 917–924.

Yang, B. and C. E. Wyman (2004). Effect of xylan and lignin removal by batch and flowthrough pretreatment on the enzymatic digestibility of corn stover cellulose. *Biotechnol Bioeng* **86**(1): 88–98.

Yang, B. and C. E. Wyman (2008). Pretreatment: The key to unlocking low-cost cellulosic ethanol. *Biofuels Bioprod Biorefin* **2**(1): 26–40.

Yang, C., Z. Shen, G. Yu and J. Wang (2008). Effect and after effect of gamma radiation pretreatment on enzymatic hydrolysis of wheat straw. *Bioresour Technol* **99**(14): 6240–6245.

Yang, C.-Y., I.-C. Sheih and T. J. Fang (2012). Fermentation of rice hull by *Aspergillus japonicus* under ultrasonic pretreatment. *Ultrason Sonochem* **19**: 687–691.

Yang, D., L.-X. Zhong, T.-Q. Yuan, X.-W. Peng and R.-C. Sun (2013). Studies on the structural characterization of lignin, hemicelluloses and cellulose fractionated by ionic liquid followed by alkaline extraction from bamboo. *Ind Crops Prod* **43**: 141–149.

Yang, G., J. R. Rich, M. Gilbert, W. W. Wakarchuk, Y. Feng and S. G. Withers (2010). Fluorescence activated cell sorting as a general ultra-high-throughput screening method for directed evolution of glycosyltransferases. *J Am Chem Soc* **132**: 10570–10577.

Yaniv, O., E. Morag, I. Borovok, E. A. Bayer, R. Lamed, F. Frolow and L. J. Shimon (2013). Structure of a family 3a carbohydrate-binding module from the cellulosomal scaffoldin CipA of *Clostridium thermocellum* with flanking linkers: Implications for cellulosome structure. *Acta Crystallogr Sect F Struct Biol Cryst Commun* **69**(Pt 7): 733–737.

Yapo, B. M. (2011). Pectic substances: From simple pectic polysaccharides to complex pectins—A new hypothetical model. *Carbohydr Polym* **86**(2): 373–385.

Yapo, B. M. and K. L. Koffi (2014). Extraction and characterization of highly gelling low methoxy pectin from cashew apple pomace. *Foods* **3**: 1–12.

Yaron, S., E. Morag, E. A. Bayer, R. Lamed and Y. Shoham (1995). Expression, purification and subunit-binding properties of cohesins 2 and 3 of the *Clostridium thermocellum* cellulosome. *FEBS Lett* **360**(2): 121–124.

Yasuda, M., K. Takeo, T. Matsumoto, T. Shiragami, K. Sugamoto, Y.-I. Matsushita and Y. Ishii (2013). Effectiveness of lignin-removal in simultaneous saccharification and fermentation for ethanol production from napiergrass, rice straw, silvergrass, and bamboo with different lignin-contents. In *Sustainable Degradation of Lignocellulosic Biomass—Techniques, Applications and Commercialization*, A. K. Chandel and S. S. da Silva (Eds.). Rijeka, Croatia: InTech-Open Science Open Minds.

Yoon, L. W., G. C. Ngoh, A. S. M. Chua and M. A. Hashim (2011). Comparison of ionic liquid, acid and alkali pretreatments for sugarcane bagasse enzymatic saccharification. *J Chem Technol Biotechnol* **86**(10): 1342–1348.

Yoon, M., J.-I. Choi, J.-W. Lee, D.-H. Park (2012). Improvement of saccharification process for bioethanol productio from *Undaria* sp. by gamma irradiation. *Radiat Phys Chem* **81**: 999–1002.

You, L. and F. H. Arnold (1996). Directed evolution of subtilisin E in *Bacillus subtilis* to enhance total activity in aqueous dimethylformamide. *Protein Eng* **9**(1): 77–83.

Yu, C.-C., T. Hill, D. H. Kwan, H.-M. Chen, C.-C. Lin, W. Wakarchuk and S. G. Withers (2014). A plate-based high-throughput activity assay for polysialyltransferase from *Neisseria meningitidis*. *Anal Biochem* **444**: 67–74.

Yu, H., G. Guo, X. Z. Keliang Yan and C. Xu (2009). The effect of biological pretreatment with the selective white-rot fungus *Echinodontium taxodii* on enzymatic hydrolysis of softwoods and hardwoods. *Bioresour Technol* **100**: 5170–5175.

Yu, L. and A. J. Mort (1996). Partial characterization of xylogalacturonans from cell walls of ripe watermelon fruit: Inhibition of endopolygalacturonase activity by xylosylation. *Progress Biotechnol* **14**: 79–88.

Yuan, K. L. and S. L. Boa (1979). Purification and characterization of endopolygalacturonase from Rhizopus arrhizus. *J Food Sci* **43**: 721–726.

Zaccolo, M., D. M. Williams, D. M. Brown and E. Gherardi (1996). An approach to random mutagenesis of DNA using mixtures of triphosphate derivatives of nucleoside analogues. *J Mol Biol* **255**(4): 589–603.

Zaidel, D. N. A. and A. S. Meyer (2012). Biocatalytic cross-linking of pectic polysaccharides for designed food functionality: Structures, mechanisms, and reactions. *Biocatal Agric Biotechnol* **1**(3): 207–219.

Zakrzewska, M. E., E. Bogel-Łukasik and R. Bogel-Łukasik (2010). Solubility of carbohydrates in ionic liquids. *Energy Fuels* **24**(2): 737–745.

Zambare, V., A. Zambare, K. Muthukumarappan and L. P. Christopher (2011). Biochemical characterization of thermophilic lignocellulose degrading enzymes and their potential for biomass bioprocessing. *Int J Energy Environ* **2**(1): 99–112.

Zamocky, M., R. Ludwig, C. Peterbauer, B. M. Hallberg, C. Divne, P. Nicholls and D. Haltrich (2006). Cellobiose dehydrogenase: A flavocytochrome from wood-degrading, phytopathogenic and saprotropic fungi. *Curr Protein Pept Sci* **7**(3): 255–280.

Zamora, R. and J. A. Sanchez Crispin (1995). Production of an acid extract of rice straw. *Acta Cient Venez* **46**(2): 135–139.

Zandleven, J., G. Beldman, M. Bosveld, J. Benen and A. Voragen (2005). Mode of action of xylogalacturonan hydrolase towards xylogalacturonan and xylogalacturonan oligosaccharides. *Biochem J* **387**(3): 719–725.

Zandleven, J., G. Beldman, M. Bosveld, H. A. Schols and A. G. J. Voragen (2006). Enzymatic degradation studies of xylogalacturonans from apple and potato, using xylogalacturonan hydrolase. *Carbohydr Polym* **65**(4): 495–503.

Zeilinger, S., D. Kristufek, I. Arisan-Atac, R. Hodits and C. P. Kubicek (1993). Conditions of formation, purification, and characterization of an alpha-galactosidase of *Trichoderma reesei* RUT C-30. *Appl Environ Microbiol* **59**(5): 1347–1353.

Zeng, W., G. Du, J. Chen, J. Li and J. Zhou (2015). A high-throughput screening procedure for enhancing α-ketoglutaric acid production in *Yarrowia lipolytica* by random mutagenesis. *Process Biochem* **50**: 1516–1522.

Zevenhoven, M. (2000). The prediction of deposit formation in combustion and gasification of biomass fuels—Fractionation and thermodynamic multiphase multi-component equilibrium (TPCE) calculations. In S. Haefele (Ed.), *Combustion and Materials Chemistry*, p. 38. Lemminkäinengatan, Finland.

Zhan, X., D. Wang, S. R. Bean, X. Mo, X. S. Sun and D. Boyle (2006). Ethanol production from supercritical-fluid-extrusion cooked sorghum. *Ind Crops Prod* **23**: 304–310.

Zhang, D., X. Chen, J. Chi, J. Feng, Q. Wu and D. Zhu (2015). Semi–Rational engineering a carbonyl reductase for the enantioselective reduction of β-amino ketones. *ACS Catal* **5**(4): 2452–2457.

Zhang, M., Z. Jiang, L. Li and P. Katrolia (2009). Biochemical characterization of a recombinant thermostable β-mannosidase from *Thermotoga maritima* with transglycosidase activity. *J Mol Catal B-Enzym* **60**(3–4): 119–124.

Zhang, Q., J. Yang, K. Liang, L. Feng, S. Li, J. Wan, X. Xu, G. Yang, D. Liu and S. Yang (2008). Binding interaction analysis of the active site and its inhibitors for neuraminidase (N1 subtype) of human influenza virus by the integration of molecular docking, FMO calculation and 3D-QSAR CoMFA modeling. *J Chem Inf Model* **48**(9): 1802–1812.

Zhang, S., D. C. Irwin and D. B. Wilson (2000). Site-directed mutation of noncatalytic residues of *Thermobifida fusca* exocellulase Cel6B. *Eur J Biochem* **267**(11): 3101–3115.

Zhang, Y., L. Wang and H. Chen (2013). Formation kinetics of potential fermentation inhibitors in a steam explosion process of corn straw. *Appl Biochem Biotechnol* **169**(2): 359–367.

Zhang, Y. H., J. Cui, L. R. Lynd and L. R. Kuang (2006a). A transition from cellulose swelling to cellulose dissolution by o-phosphoric acid: Evidence from enzymatic hydrolysis and supramolecular structure. *Biomacromolecules* **7**(2): 644–648.

Zhang, Y. H. and L. R. Lynd (2006). A functionally based model for hydrolysis of cellulose by fungal cellulase. *Biotechnol Bioeng* **94**(5): 888–898.

Zhang, Y. H. P. (2008). Reviving the carbohydrate economy via multi-product lignocellulose biorefineries. *J Ind Microbiol Biotechnol* **35**(5): 367–375.

Zhang, Y. H. P., M. E. Himmel and J. R. Mielenz (2006b). Outlook for cellulase improvement: Screening and selection strategies. *Biotechnol Adv* **24**(5): 452–481.

Zhang, Y.-H. P., S.-Y. Ding, J. R. Mielenz, J.-B. Cui, R. T. Elander, M. Laser, M. E. Himmel, J. R. McMillan and L. R. Lynd (2007). Fractionating recalcitrant lignocellulose at modest reaction conditions. *Biotechnol Bioeng* **97**(2): 214–223.

Zhang, Y.-H. P. and L. R. Lynd (2004). Toward an aggregated understanding of enzymatic hydrolysis of cellulose: Noncomplexed cellulase systems. *Biotechnol Bioeng* **88**(7): 797–824.

Zhao, G., E. Ali, M. Sakka, T. Kimura and K. Sakka (2006a). Binding of S-layer homology modules from *Clostridium thermocellum* SdbA to peptidoglycans. *Appl Microbiol Biotechnol* **70**(4): 464–469.

Zhao, G., H. Li, B. Wamalwa, M. Sakka, T. Kimura and K. Sakka (2006b). Different binding specificities of S-layer homology modules from *Clostridium thermocellum* AncA, Slp1, and Slp2. *Biosci Biotechnol Biochem* **70**(7): 1636–1641.

Zhao, X., K. Cheng and D. Liu (2009). Organosolv pretreatment of lignocellulosic biomass for enzymatic hydrolysis. *Appl Microbiol Biotechnol* **82**(5): 815–827.

Zhao, X., L. Zhang and D. Liu (2012). Biomass recalcitrance. Part I: The chemical compositions and physical structures affecting the enzymatic hydrolysis of lignocellulose. *Biofuels Bioprod Biorefin* **6**(4): 465–482.

Zhao, Y., B. Wu, B. Yan and P. Gao (2004). Mechanism of cellobiose inhibition in cellulose hydrolysis by cellobiohydrolase. *Sci China Ser C: Life Sci* **47**(1): 18–24.

Zheng, F. and S. Ding (2013). Processivity and enzymatic mode of a glycoside hydrolase family 5 endoglucanase from *Volvariella volvacea*. *Appl Environ Microbiol* **79**(3): 989–996.

Zheng, H., Y. Liu, M. Sun, Y. Han, J. Wang, J. Sun and F. Lu (2014). Improvement of alkali stability and thermostability of *Paenibacillus campinasensis* Family-11 xylanase by directed evolution and site-directed mutagenesis. *J Ind Microbiol Biotechnol* **41**(1): 153–162.

Zheng, Y., H. Lin and G. T. Tsao (1998). Pretreatment for cellulose hydrolysis by carbon dioxide explosion. *Biotechnol Prog* **14**(6): 890–896.

Zheng, Y., H.-M. Lin, J. Wen, N. Cao, X. Yu and G. Tsao (1995). Supercritical carbon dioxide explosion as a pretreatment for cellulose hydrolysis. *Biotechnol Lett* **17**(8): 845–850.

Zhou, C., J. Ye, Y. Xue and Y. Ma (2015). Directed evolution and structural analysis of alkaline pectate lyase from alkaliphilic *Bacillus* sp. N16-5 for improvement of thermostability for efficient ramie degumming. *Appl Environ Microbiol* **81**: 5714–5723.

Zhou, J., Y.-H. Wang, J. Chu, L.-Z. Luo, Y.-P. Zhuang and S.-L. Zhang (2009). Optimization of cellulase mixture for efficient hydrolysis of steam-exploded corn stover by statistically designed experiments. *Bioresour Technol* **100**(2): 819–825.

Zhu, J. Y., G. S. Wang, X. J. Pan and R. Gleisner (2009a). Specific surface to evaluate the efficiencies of milling and pretreatment of wood for enzymatic saccharification. *Chem Eng Sci* **64**(3): 474–485.

Zhu, S., Y. Wu, Q. Chen, Z. Yu, C. Wang, S. Jin, Y. Ding and G. Wu (2006). Dissolution of cellulose with ionic liquids and its application: A mini-review. *Green Chem* **8**(4): 325–327.

Zhu, Z., N. Sathitsuksanoh, T. Vinzant, D. J. Schell, J. D. McMillan and Y. H. P. Zhang (2009b). Comparative study of corn stover pretreated by dilute acid and cellulose solvent-based lignocellulose fractionation: Enzymatic hydrolysis, supramolecular structure, and substrate accessibility. *Biotechnol Bioeng* **103**(4): 715–724.

Zong, Z., L. Gao, W. Cai, L. Yu, C. Cui, S. Chen and D. Zhang (2015). Computer-assisted rational modifications to improve the thermostability of β-glucosidase from *Penicillium piceum* H16. *BioEnergy Research* **8**: 1384.

Zverlov, V. V., K.-P. Fuchs and W. H. Schwarz (2002a). Chi18A, the endochitinase in the cellulosome of the thermophilic, cellulolytic bacterium *Clostridium thermocellum*. *Appl Environ Microbiol* **68**(6): 3176–3179.

Zverlov, V. V., N. Schantz, P. Schmitt-Kopplin and W. H. Schwarz (2005a). Two new major subunits in the cellulosome of *Clostridium thermocellum*: Xyloglucanase Xgh74A and endoxylanase Xyn10D. *Microbiology* **151**(10): 3395–3401.

Zverlov, V. V., N. Schantz and W. H. Schwarz (2005b). A major new component in the cellulosome of *Clostridium thermocellum* is a processive endo-β-1,4-glucanase producing cellotetraose. *FEMS Microbiol Lett* **249**(2): 353–358.

Zverlov, V. V., G. A. Velikodvorskaya and W. H. Schwarz (2002b). A newly described cellulosomal cellobiohydrolase, CelO, from *Clostridium thermocellum*: Investigation of the exo-mode of hydrolysis, and binding capacity to crystalline cellulose. *Microbiology* **148**(1): 247–255.

Zverlov, V. V., G. A. Velikodvorskaya and W. H. Schwarz (2003). Two new cellulosome components encoded downstream of celI in the genome of *Clostridium thermocellum*: The non-processive endoglucanase CelN and the possibly structural protein CseP. *Microbiology* **149**(2): 515–524.

Zverlov, V. V., G. V. Velikodvorskaya, W. H. Schwarz, K. Bronnenmeier, J. Kellermann and W. L. Staudenbauer (1998). Multidomain structure and cellulosomal localization of the *Clostridium thermocellum* cellobiohydrolase CbhA. *J Bacteriol* **180**(12): 3091–3099.

Zweig, A., W. G. Hodgson and W. H. Jura (1964). The oxidation of methoxybenzenes. *J Am Chem Soc* **86**: 4124–4129.

Zykwinska, A. W., M. C. Ralet, C. D. Garnier and J. F. Thibault (2005). Evidence for in vitro binding of pectin side chains to cellulose. *Plant Physiol* **139**(1): 397–407.

Index

Printed and bound by CPI Group (UK) Ltd, Croydon, CR0 4YY

24/10/2024

01778301-0003